Shifting Ground

Shifting Ground

The Changing Agricultural Soils of China and Indonesia

Peter H. Lindert

The MIT Press
Cambridge, Massachusetts
London, England

This book was set in Sabon by Wellington Graphics, Westwood, MA and printed and bound in the United States of America.

Printed on recycled paper.

Library of Congress Cataloging-in-Publication Data

Lindert, Peter H.
  Shifting ground : the changing agricultural soils of China and Indonesia / Peter H. Lindert.
    p. cm.
  Includes bibliographical references (p.).
  ISBN 0-262-12227-8 (hc. : alk. paper)
  1. Soils—China—Quality. 2. Soil degradation—China. 3. Agriculture—Environmental aspects—China. 4. Soils—Indonesia—Quality. 5. Soil Degradation—Indonesia. 6. Agriculture—Environmental aspects—Indonesia. I. Title.
S599.6.C5 L57 2000
631.4'951—dc21   00-020373

Contents

Preface

The work reported here is a group effort dating back to 1992. The initial help was, of course, the key financial support from the U.S. National Science Foundation (grants SES9213812 and SBR9512111), supplemented by a grant from the University of California's Pacific Rim Research Program that helped launch the Indonesian side of the project in 1994. My colleague Michael Singer served as a co-investigator in framing the initial proposal. Thereafter he provided frequent help in my struggles with the unfamiliar field of soil science. Although he has reduced the number of my errors, he and the others mentioned here should not be held accountable for any errors that remain.

The data gathering for China took place between early 1993 and early 1995. The first step was the gathering of the 1932–1944 set of soil profiles from rare published materials in the United States. This work in 1993 relied mainly on Joann Lu's able research assistance, though the U.S. National Agricultural Library also helped by lending its copy of the rare volumes 20–24 of the *Soil Bulletins*. The next data set, covering 1950–1964, came from three main sources in 1993–1995. First, I translated and processed profile data from Viktor Kovda's 1959 book, *Ocherki Prirody i Pochv Kitaia* (Notes on the Nature and Soils of China). Then Professor Chen Ziming was extremely helpful in obtaining a photocopy of the rare publication of the First National Soil Survey (1958–1961). Later Wu Wanli added profiles from an in-depth 1961–1964 study of the soils of Nei Mongol.

The 1980s data for China drew on the large Second National Soil Survey data originally gathered in 1981–1986. Gathering these data was not easy, however, during a 1993 visit to Beijing and Nanjing. At that time the various province volumes were still in press or in manuscript.

Fortunately, the Institute of Soil Science, Academia Sinica, was able to direct me to the necessary soil profile data. For some provinces, this meant purchasing the newly published volumes that were appearing in 1994. For other provinces, it meant photocopying copies of soil profile data from province volumes that were still in press. Although one could now amass a fuller data set for the 1980s than I could gather in 1993–1995, the data set used in this volume suffices for most purposes.

Back in Davis, the data for China were processed by Joann Lu between 1993 and 1995. This involved not only translations and data entry, but also collating locations with maps, a task complicated by the changes in China's provincial and county boundaries. I thank Joann for this work, and also Wu Wanli for helping us judge some difficult soil and geographical data. G. William Skinner and Zhang Bing also helped by supplying, and helping to correct, the county-level data on agriculture in 1985.

So important were the roles of Joann Lu and Wu Wanli that they became coauthors with me of two 1996 articles on China's soil changes, one in *Soil Science* and one in the *Journal of Environmental Quality* (see the references for the full citations).

The Indonesian side of the project was born suddenly in late 1993. Since proposing the overall project in 1992, I had been casting about for a second country to supplement the experience of China. The original grant proposal had targeted India, but an Indian expedition to gather soil data looked only moderately promising, both because abundant soil data started only in the 1950s and because the relevant experts in Nagpur were too busy to offer extensive collaboration. Then in late 1993 I received a review copy of S. M. Chin A Tam's newly published *Bibliography of Soil Science in Indonesia, 1890–1963*. This made it clear that there were at least a few hundred good soil profiles for Indonesia from before 1963, most of them available at the Center for Soil and Agroclimate Research (CSAR) in Bogor. By luck, my friend Jeanine Pfeiffer was then in Bogor doing agronomic and botanical research at Kebun Raya Indonesia. She confirmed that it should be possible to find some such data at CSAR, and a joint proposal was quickly submitted to the University of California's Pacific Rim Research Program, which kindly backed our gamble. By April 1994 Jeanine's detective work had turned up a far richer data set than I had envisioned from reading Chin A Tam's book. A storeroom at CSAR housed an uncatalogued archive of more than 100,000 soil profile analyses from 1923 to 1983. She secured the cooperation of Dr. Hardjosubroto Subagjo and others at CSAR for gathering and copying the soil profile data. Throughout 1994–1995, but particularly during my work visit to Bogor in October 1994, Jeanine was the master organizer of the links with CSAR and the data-collating process there. She also clarified the translation of agronomic and botanical terminology from Indonesian into English. The data gathering and copying was aided by CSAR's provision of space and the extensive labors of several staffers there, particularly Atiek Widayati and Diah Setyorini. After the 1994 visit, Atiek Widayati continued to help with data transmission, and UC-Davis students Yenny Leswati and Amelia Salem entered the data at the Davis end.

The task of finding the agricultural effects of Indonesia's soil characteristics in chapter 8 required a careful edited database for Indonesia. I thank Pierre van der Eng for supplying not only computer files of extensive official data by district, but also wise warnings on how to handle those data.

Once the results were analyzed, paper drafts began to appear, first for China and then for Indonesia. The original hope was that two books could be written and edited by separate teams of authors for China and for Indonesia. Yet the China and Indonesia collaborators were unable to find time from other duties to write their chapters, the exceptions being partial chapter drafts kindly written by Ren Jianhuang for China and Pierre van der Eng for Indonesia. Although this forced a retreat to the Lindert-Lu-Wu articles plus this single-authored book, team members continued to supply helpful comments and suggestions, as did other scholars who read drafts and heard presentations. I thank J. Sri Adiningsih, Taco Bottema, Colin Carter, Kenneth G. Cassman, Chen Ziming, Pierre Crosson, Erle C. Ellis, Gong Zitong, Li Chingkwei, Jeanine Pfeiffer, Scott Rozelle, Djoko Santoso, Shi Xuezheng, Chessy Q. Si, Michael Singer, G. William Skinner, H. Subagjo, M. Sudjadi, C. Peter Timmer, Thomas Tomich, Pierre van der Eng, Meine van Noordwijk, Xi Chengfan, Wang Siming, Jeffrey G. Williamson, and Wu Wanli for their cooperation and advice. Helpful comments were also offered by anonymous referees and by seminar participants at the All-UC Group in Economic History, the Association for Asian Studies 1996 meetings, the Australian

National University, the University of Adelaide, the University of California–Davis, UCLA, the University of Illinois, Harvard University, the Center for International Forestry Research (Bogor), Melbourne University, the Millennium Institute, Northwestern University, and Stanford University.

The author and publisher gratefully acknowledge the following permissions to reproduce parts of these three publications:

1. Williams & Wilkins, for permission to draw from Peter H. Lindert, Joann Lu, and Wu Wanli, "Trends in the Soil Chemistry of South China since the 1930s," *Soil Science* 161 (May 1996): 329–42.

2. The *Journal of Environmental Quality*, for permission to draw from Peter H. Lindert, Joann Lu, and Wu Wanli, "Trends in the Soil Chemistry of North China since the 1930s," *Journal of Environmental Quality* 25, no. 4 (November–December 1996): 1168–1178.

3. *Economic Development and Cultural Change*, for permission to draw from Peter Lindert, "The Bad Earth? China's Soils and Agricultural Development since the 1930s," *Economic Development and Cultural Change* 47, 4 (July 1999), pp. 701–736.

The quantitative soil histories of China and Indonesia should continue, especially now that both countries are building up large computerized compilations from the 1980s on. To request the CD-ROM of the full data sets offered in this study, readers should request "Soil Data Sets for China and Indonesia," either from the Inter-University Consortium on Political and Social Research (ICPSR) or from the author:

Peter H. Lindert, Agricultural History Center
University of California–Davis
One Shields Avenue
Davis, CA 95616 USA
phlindert@ucdavis.edu

Those interested in a full version of the China data set should also ask for a separate transmission or mailing of the Excel 5.0 file "Soil Data '80s Hubei," which was inadvertently not included on the compact disk. (A copy of the Hubei-1980s file has already been sent to several collaborators in China.) The CD files and the Hubei file can also be downloaded from http://aghistory.ucdavis.edu

I
Judging Soil Trends

1

Current Concerns

Alarm Bells

If the world is losing its productive agricultural soils, the damage is likely to be most severe in the desocializing and developing countries of the Second and Third Worlds. Any deterioration of food supply presents a greater threat to poorer countries. These are also the countries in which soil degradation should be most expected, since they are less able to wait for the delayed benefits from investments in soil and forest conservation, in fertilizer, and in water control. Is soil degradation accelerating in developing countries? How much damage is it likely to cause? What countermeasures are needed?

Two opposing majority views seem to prevail on the issue of soil degradation. One is the majority of writers on this subject, and the other is the majority of all people, most of them silent on this issue. As with so many social issues, the majority among those who write on the subject are alarmed (e.g., Eckholm 1976; Dregne 1982, 1992; Smil 1984; World Resources Institute 1988–1994; Orleans 1992; Brown 1995; Gardner 1996). By contrast, the likely view of the silent majority is that soil degradation is only a secondary concern. That the more outspoken side of the debate tends toward pessimism seems natural enough: The incentive to commit time and reputation to writing on a particular issue is stronger for those who feel that society needs to be warned about that issue.

The main international institutions have taken the threat of soil degradation seriously in the 1980s and 1990s. Development agencies are increasingly committed to monitoring gains and losses in natural resource assets. The United Nations and the World Bank have set guidelines for incorporating environmental accounts into the traditional measures of national product (Ahmed, El Serafy, and Lutz, 1989). Although environmental investments and depreciation are always harder to quantify than conventional market products, there is no denying that our view of economic progress is in danger of either overoptimism or overpessimism if it fails to measure changes in the stock of natural resources. The usual fear about our conventional accounts is that they are too optimistic, since conventional gross domestic product (GDP) measures fail to deduct values for the depreciation of a vital natural resource. There is a growing global awareness that resource depletion needs to be measured and priced. It must also be combated, sooner or later, because many natural resources cannot be depleted indefinitely without canceling the gains from economic growth.

Some writers have bravely produced summary estimates of what soil degradation and related environmental damages have cost some of the developing countries. Table 1.1 presents the leading cost estimates. The numbers are disturbingly large. As a share of gross national product these costs look larger than even most estimates of the costs of developing countries' corruption or their barriers to international trade. Given the enormity of the GNP denominator, not many estimates of the cost of anything would exceed 15 percent, as do some of these land degradation estimates. Even America's Great Depression cost only a third of GNP in its worst year, and the land degradation costs in table 1.1 could potentially last longer than that depression. Behind such suggestive cost figures lie other writers' appraisals of physical damage, appraisals that we must weigh carefully in this and the next chapter.

The concern over soil degradation stems from three related beliefs about the physical processes at work:

1. that erosion is the main source of soil degradation;
2. that the degradation is accelerating; and
3. that humans are responsible.

Although soil and other land resources can be degraded in many different ways, the warnings in the literature on soil trends focus on erosion above all, with only secondary attention to soil nutrient losses from intensive cropping and inadequate fertilization.

Table 1.1
Estimates of national environmental damages from land degradation in selected developing countries in the 1980s

Country	Types of environmental damage	Year	Annual damage as a percent of GNP
Burkina Faso	Crop, livestock, and fuelwood losses due to land degradation	1998	8.8
Costa Rica	Coastal fisheries destruction, deforestation, and soil erosion	1989	7.7
Indonesia	Soil erosion and deforestation	1984	4.0
Madagascar	Land burning and soil erosion	1988	5.0–15.0
Malawi	Soil erosion and deforestation	1988	2.8–15.2
Nigeria	Soil degradation, deforestation, water pollution, other erosion	1989	17.4

Sources: Barbier and Bishop 1995, p. 133, citing several other studies. Barbier (1996, p. 1) refers to these estimates as "often more illustrative than definitive, due to paucity of empirical data and various methodological problems." Additional estimates for Africa are summarized in Scherr 1999, p. 28.

Yet we have almost no direct data on what we have done to the land, or when and where we have done it. There is no set of usable numbers for following the trends in soil quality or soil endowment over the decades of this century, not even for the United States and the other industrialized countries. As we shall see, what have been offered as data on soil trends are a wide variety of indirect clues, not a true quantitative history. This information gap should be filled as soon as possible, especially for the developing countries where the dangers seem most immediate.

This book presents a tentative history of agricultural soils in two key developing countries that have always been at the center of the debate over land degradation. China and Indonesia comprise large shares of world resources and activities relevant to land degradation. As shown in table 1.2, they represent an eighth of all "degraded lands" as estimated by the World Resources Institute. If human population and intensive agriculture are seen as the main threat to the land, then China and Indonesia loom even larger, with a quarter of the world's population and more than two-fifths of its agricultural labor force. Comparing the estimates of degraded lands to all cultivated fields and all forests suggests that China's

Table 1.2
Approximate shares of China and Indonesia in world resources and activities relevant to land degradation, 1990–1991

Shares of the world's	China	Indonesia	Both
Population 1991	21.5%	3.4%	24.9%
Agricultural labor 1990	40.4%	3.7%	44.1%
Fields and forests 1991	7.2%	1.6 %	8.8%
"Degraded lands" 1990	11.7%	0.8%	12.5%

Notes: Population figures for mid-1991 are from World Bank 1993, pp. 238–39. Agricultural labor force estimates are from World Bank 1997, pp. 220–21, interpolating to these 1990 estimates: China, 503.4 million; Indonesia, 45.6 million; world, 1,244.7 million. Such figures for China typically overcount the share of labor devoted to agriculture by counting all rural population as agricultural households. The fields and forests areas are the sum of all arable and permanent cropland plus permanent pasture and forests and woodlands, from Food and Agriculture Organization 1995, table 1, pp. 3–9.

The "degraded lands" figures are from three sources, with variation in the measure of degraded land area. For China, the series published by the Ministry of Water Resources and Electrical Power cited by Huang and Rozelle (1995, p. 856) estimates about 135 million hectares of "erosion area" and about 7.5 million of "area affected by salinization." These were counted as China's degraded lands here, though the Ministry's additional 31 million hectares of "flood and drought area" was excluded on the grounds of being less comparable with the worldwide estimates. Indonesia's degraded lands are those 9.667 million hectares of degraded lands estimated by Buro Pusat Statistik for April 1987, as cited in table 2.3. The world estimates are the 1,215.4 hectares of "moderate, severe, and extreme" human-induced soil degradation as stated by World Resources Institute (1992, p. 112) on the basis of the GLASOD map discussed in chapter 2.

land degradation is worse than the world average, but that Indonesia's has not (yet) reached the world average.

The geography and history of China and Indonesia also recommend these two countries as key test cases. China has always been at the center of global concerns about desertification and water erosion. Deserts, dust, and drought threaten China's northwest and heavy rains and floods threaten the rest of the country. Twentieth-century China is also famous for the bad and good that humans may have done to the soil. The country may have accelerated erosion and soil exhaustion with intense agriculture, with overemphasis on expanding grain area, and by stripping already meager forests in the Great Leap's all-out campaign for smelting steel in every village. On the other hand, even pessimists about global soil losses have praised China's campaigns for afforestation (e.g., Eckholm 1976, chap. 2). China has also has an impressive history of investments in water control, especially during the Mao years. Still, land degradation in China causes much alarm because it is believed to have destroyed much of China farmlands. Vaclav Smil (1984, pp. 67–70) most clearly sounded the alarm bell about losses of China's cultivated area since 1957:

China lost, to urban and rural construction and to natural disasters, . . . 33.33 million ha, an incredible 29.8 percent of the 1957 total [cultivated area]. This is a truly frightening figure . . . the world's most populous nation with already scarce farmland losing nearly a third of its prime cropland in just one generation! The losses were partially made up by the . . . reclamation campaigns . . . so the net loss was 12 million ha, or 10.73 percent of the 1957 total. Even . . . ignoring the inferior quality of most of the reclaimed land—it is an awesome drop.

The accelerating pace of human intervention is believed to be the main culprit, acting mainly through earlier deforestation, but also through urbanization, poor water control, and excessive cropping.[1]

China has already suffered at least one spectacular policy failure caused by underestimating the importance of erosion on the loess plateau. The infamous Sanmenxia dam and reservoir was built across the Huang He in the Great Leap era in the hope of supplying large amounts of hydroelectric power while also helping irrigation and flood control. The heroic decision to build Sanmenxia was based partly on the belief that a system of check dams and massive revegetation on the loess plateau could prevent siltation, which proved incorrect. Between 1960 and 1964 the Sanmenxia lost half its generating capacity to silt (Smil 1987, p. 216), and much of the dam is locked open by siltation problems to this day, leaving a massive monument to resource waste. It remains to be seen whether the Three Gorges Dam or the Yangzi will succumb to siltation, as some have feared.

Indonesia is also a leading case study in soil dynamics. Volcanoes and erosion created most of the country's lands, and erosion continues to extend river deltas at the expense of the uplands. A 1997 crisis spotlighted Indonesia as a country where environmental controls had collapsed. Fires started mainly to clear forests for cash crops polluted the air for months on end, food was scarce on Java, and a desperate population

on the outer islands grabbed and sold rain forest resources at an accelerated pace. Even before this crisis, the country's rapid deforestation and
shifting slash-and-burn agriculture had been singled out for international
censure. In Indonesia, as in China, the evidence of human impact on the
soil should be particularly visible.

Concerns about erosion also have a long history in Indonesia. In 1866
K. F. Holle warned that soil erosion in the Indonesian archipelago was a
great danger that was creeping ever nearer.[2] The concern continues.
Many have recently expressed fears that what was then a creep has
become a gallop, as accelerating human population growth and increasingly intense cultivation undermine Indonesia's ability to feed itself (Donner 1987; Carson 1989; Repetto et al. 1989; Soemarwoto, 1991).
Authors have buttressed this concern with a wide range of indirect
evidence. Forests are being stripped, rivers are muddy, river deltas have
pushed out to sea, some farm plots have been lost to gullies, and experiments with steep slopes and high rainfall show rapid rates of gross soil
loss. The fears about Indonesia's erosion are greatest for the crowded
island of Java:

> In the mountains of . . . Java . . . the connection between population density and
> environmental degradation is unmistakable. . . . At least one third of Java's
> cultivated mountainous areas are eroding seriously. More than 1 million hectares
> of these holdings are now useless for farming, depriving people of their only
> means of support. Excessive erosion directly threatens the livelihood of about 12
> million poor in Java.[3]

What Soil Degradation Means

To set the stage for the chapters ahead, it is important to understand the
meaning of soil degradation and its main types. *Soil degradation* refers to
any chemical, physical, or biological change in the soil's condition that
lowers its agricultural productivity, defined as its contribution to the
economic value of yields per unit of land area, holding other agricultural
inputs the same. A synonym for the soil's "agricultural productivity" is
soil "quality."

Table 1.3 lists the main types of soil degradation. Most can be either
human induced or natural. The order in which they are listed anticipates
some of this book's conclusions. First come the varieties of chemical

Table 1.3
Main types of soil degradation

Type of soil degradation	Its manifestations in the topsoil
Chemical deterioration	
Loss of soil nutrients and organic matter	Lower levels of these
Soil alkalization, salinization, and/or sodicity	Rises in pH above 8.0, in salt content, and in sodium ion content
Soil acidification	Declines in pH below 6.0
Water and wind erosion	Shallower topsoil layer (the "A horizon"), *and* loss of macro-nutrients (nitrogen, phosphorus, potassium, sulfur) in the remaining topsoil

Notes: Some other types of soil degradation not pursued in this book are soil pollution and terrain deformation due to water or wind erosion. These are likely to have brought less damage than the chemical and erosion types listed above.

deterioration, each of which can result from human agricultural misuse of the soil. Each also lends itself directly to measurement here and shows a striking set of patterns in China and Indonesia since the 1930s.

Soil erosion by water and wind is usually thought to be the greatest threat, as in the alarms sounded above for China and Indonesia. Yet as we will see, the evidence suggests that water and wind erosion is not the dominant threat to soil quality on the cultivated land areas of these countries. Erosion is a visible reality mostly in areas where it has been going on since before human settlement or cultivation, without being human induced. So say the two kinds of indirect evidence about erosion trends listed in table 1.3: both trends in the thickness of the topsoil layer and trends in the levels of the key nutrients in the more erosion prone areas.

The Importance of Soil Chemistry

Assessing either kind of soil degradation, either chemical deterioration or soil erosion, requires focusing on changes in the quality of the topsoil layer. Many of the soil characteristics that matter most to yields are just chemical characteristics, introduced in table 1.4 and figure 1.1, that have

Table 1.4
Some agricultural impacts of topsoil chemistry

Soil chemistry characteristic	Its effect on plant growth and crop yields	Yield effect interacts	Natural geographic pattern	Ways humans control this soil characteristic
pH (hydrogen ion activity)	Controls all chemical reactions governing plant growth	with soil OM	Cold, dry climates → high pH (alkalinity) Hot, wet climates → low pH (acidity)	Improve water control to fight alkalinity, apply lime to fight acidity
Organic matter (OM) and total nitrogen (N)	Decay of OM supplies energy and materials for soil microorganisms that contribute to plant growth. N is key to building plant proteins, to carbohydrate use, and to plants' uptake of other nutrients. The two levels are related, with the OM/N ratios cycling widely around 20:1.	with soil pH, K, soil texture, and moisture (and OM and N interact with each other)	Cold, moist climates → high OM, N, (OM/N) Hot, dry climates → low OM, N, (OM/N)	Apply fertilizers, both synthetic and organic (manure, green waste); choose nitrogen-fixing crops; control crop removals
Total phosphorus (P)	P in a form "available" for plants is essential for photosynthesis, N-fixation, and crop maturation. Most natural P is released into available forms only very slowly.	with soil pH, texture	Cold, dry climates → high availability of P Hot, wet climates → low availability of P	Apply fertilizers, control crop removals

Table 1.4
(continued)

Soil chemistry characteristic	Its effect on plant growth and crop yields	Yield effect interacts	Natural geographic pattern	Ways humans control this soil characteristic
Total potassium (K)	K in a form "available" for plants activates enzymes for plant metabolism, starch synthesis, nitrate reduction, and sugar degradation. Most natural K is released into available forms only very slowly.	with soil OM, pH, texture	Cold, dry climates → high K; Hot, wet climates → low K	Apply fertilizer, control leaching, control crop removals

Notes: For overviews of the underlying processes, see Brady 1990, Chaps. 8–12, and China Institute of Soil Science 1990, Chaps. 19, 24, 25, and 29–31.

Omitted here, for want of sufficient historical data from China and Indonesia, are these additional soil characteristics:

- physical composition and compaction
- sulfur, the fourth main macronutrient
- micronutrients or trace elements (iron, manganese, zinc, copper, boron, molybdenum, cobalt, and chlorine), an insufficiency or toxic abundance of which can inhibit plant growth.

Also omitted is another key parameter, soil texture, which divides the soil into particle sizes, summarized by the shares of sand versus silt versus clay. Texture interacts in nonlinear ways with the characteristics featured here, as suggested in the fourth column. Most of the historical data sets used in this book include texture, and all statistical results have controlled for its nonlinear influences.

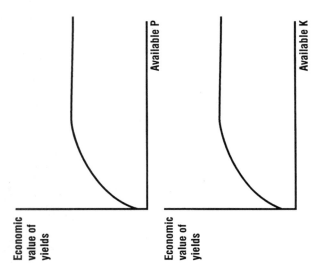

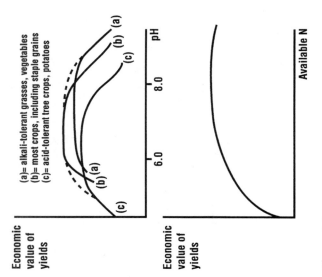

Figure 1.1
Some agricultural impacts of topsoil chemistry

been measured since at least the 1930s. They are not the only characteristics that determine soil productivity, but they are leading determinants nonetheless (Doran et al. 1984).

The first key soil characteristic is pH, defined as the negative logarithm of the hydrogen ion concentration. Acidic (pH below 6.0, say) or alkaline (pH above 8.0) soils generally yield less economic value than neutral soils (pH between 6.0 and 8.0). The agricultural cost of acidic or alkaline soils depends on the crop choice, as the schematic curves in the first panel of the figure suggest. Most crops, including wheat and maize, grow best in the pH-neutral range, as sketched in curve (b). Some crops such as tree crops and potatoes and, to a lesser extent, rice, can tolerate acid conditions. These tend to be crops that grow well in wet conditions. On the other hand these acid-tolerant crops, like most of the staple foods, do poorly under alkaline conditions (high pH). Alkalinity here is a convenient correlate of two other soil problems that are common in harsh, dry climates: salinity (high salt buildup in the topsoil) and sodicity (high sodium ion concentrations). Extracting economic value from alkaline and saline soils requires choosing crops that can tolerate high pH, high salt, and high sodium, such as barley, certain fodder grasses, and sugar beet.

Free to choose among crops in response to the pH of the soil, agricultural systems tend to press up against the dashed envelope curve sketched in the first panel of figure 1.1. This curve is usually discovered by local trial and error. To give the nonlinear role of soil pH its due, I shall represent it by two variables in what follows:

Acidity = any shortfall of pH below 6.0;

Alkalinity = any excess of pH above 8.0.

Beyond pH, soil productivity also depends on its nutrient levels. Soil nutrients are traditionally divided into two groupings: "macronutrients," with relative high concentrations in the soil [nitrogen (N), phosphorus (P), potassium (K), and sulfur (S)] and "micronutrients," or trace elements, particularly iron, manganese, zinc, copper, boron, molybdenum, cobalt, and chlorine. Historical data allow us to follow the levels of the first three macronutrients, the classic "NPK group." These macronutrients generally make nonlinear contributions to soil productivity. Beyond some sufficiency levels, extra P or extra K add nothing to soil productivity, and beyond some very high level having additional N can even detract from yields. Over the range that is relevant for China and Indonesia, however, more of each macronutrient is better. Note the distinction in table 1.4 between the total levels of N or P or K and their available levels. Plant growth responds only to "available" N or P or K, the small shares of the total endowments that are in the right chemical forms for plant growth reactions. Our historical data, however, report the total levels. These must be considered very indirect measures of the macronutrients, especially in the case of P and K, most of the total endowment of which is released only very slowly into available forms.

These few soil characteristics suffice as major clues to soil degradation and its effects on crop yields. They are good indicators not only of the main kinds of chemical degradation but also of soil erosion. Erosion does more on the site than just make the soil layer shallower. Erosion also removes key soil nutrients in ways that deprive the remaining root zone of soil nutrients (Frye, Bennett, and Buntley 1985, pp. 339–341; Larson, Pierce, and Dowdy 1983; El-Swaify, Moldenhauer, and Lo 1985, chaps. 22, 23, 28, and 29). Experiments in the United States have found that the topsoil lost to wind and water carries away five times as much organic matter and available nitrogen, three times as much available phosphorus, and twice as much available potassium as remains in the topsoil left behind after erosion (Brady 1990, p. 434).

The key effects of erosion on the N, P, and K of the remaining soil have been used to guess at the nutrient losses that erosion has meant for the soils of China. Wen Dazhong (1993, pp. 69–70) has made conjectures on losses of soil nutrients in the topsoil layers of three erosion areas in China and for the nation as a whole:

• On the loess plateau, soil "organic matter of the topsoil of some plateau areas has decreased from 3 to 0.3%, total nitrogen content from 0.2–0.3 to 0.03%, and available phosphorus content from 50 to 5 ppm by erosion." About 30 percent of the loss was from cultivated lands.

• In the south, "organic matter, nitrogen, and phosphorus in the topsoil in the eroded region have been reduced to 10%, 5%, and 2% of their original contents in the soils." Given that most of the south is both cultivated and on the "eroded" map, the implication is that something like this is lost on cultivated areas throughout the south.

• In the northeast rolling hills region, "[s]ince the land was first cultivated, the amounts of organic matter and total N in topsoils on sloping
cropland in the region have been reduced from 11% and 0.6% to 2–3%
and 0.2%, respectively."

• For China as a whole,

it is estimated that 5500 million tonnes of soil is lost annually in China due to
erosion. This soil carries with it 27.3 million tonnes of organic matter, 5.5 million
tonnes of nitrogen, 0.06 million tonnes of available phosphorus, and 0.5 million
tonnes of available potassium from the 150 million hectares. The annual total of
N, P, and K lost equals 46%, 2%, and 63%, respectively, of the total N, P, and K
applied annually in Chinese agriculture.[4]

Such cumulative nutrient losses could have reduced China's grain yields
by 3 percent (Wen 1993, p. 70). In addition, erosion and poor water
control may have raised alkalinity/salinity in the arid North and acidity in
the humid South. Others have shared Wen's pessimism about China's
trends in soil fertility (Thorp 1939, p. 159; Dregne 1982; World Resources Institute, 1988, 1992; Oldeman, van Engelen, and Pulles 1990).

Thus on the soil it leaves behind erosion leaves its fingerprints, a set of
clues that we can use to determine where and when erosion has been a
dominant force.

The Road Ahead

Judging what we have done to the land in the twentieth century, and what
difference it makes to agriculture, requires the paving of a long road, one
firm enough to carry us to reliable conclusions. This book cannot complete such a road, but it can map it and lay out a tentative path for China
and Indonesia. The next two chapters will begin to transform the database used to map soil trends. Chapter 2 takes on the negative initial task
of showing some of the severe flaws in the current soil trend evidence,
allowing chapter 3 to begin building a better, more appropriately historical database for China and Indonesia.

For each country, the journey starts with careful estimation of how soil
chemistry has changed over the decades for given kinds of soils. Chapter
4 for China, chapter 7 for Indonesia, and the corresponding appendices
cover that long stretch of the journey. The movements in soil characteristics turn out to be varied, with some chemical characteristics

improving when others worsen. The same chapters explore the likely sources of these soil changes.

Given the emerging outlines of a soil history, the next step is to decide how to weigh the economic importance of the variety of revealed trends. Which of them have a great impact on agriculture and the economy, and which matter little? The difficult task of setting up the right real-world experiment to weigh these impacts requires a different approach in different countries, depending on their economic institutions. For China before the 1990s we can take statistical advantage of some political realities that helped make each county (*xian*) a somewhat separate experiment. Chapter 5 designs such a test for China, and chapter 6 explores the economic consequences of soil quality that the test reveals. There we can arrive at a first reckoning of what economic difference soil changes make, including changes in the extent and the quality of the lands being cultivated. For Indonesia, the same task of translating soil changes into economic effects is undertaken in chapter 8. The final chapter reaches a tentative set of destinations on several issues.

The two national soil histories yield tentative conclusions that are striking and often similar between these two countries:

On Trends in Soil Characteristics

• Intensifying agriculture does not enhance soil nitrogen and organic matter, and often depletes it. Total N and organic matter drop when previous forest or grasslands are first cleared and cultivated. The decline may continue thereafter, as cultivation intensifies, which also happened on some classic long-term field experiments.

• Unlike those of nitrogen and organic matter, the total soil endowments of phosphorus and potassium have risen in the long-cultivated regions of China and Indonesia.

• Soil alkalinity did not worsen in north China between the 1930s and the 1980s. Rather it remained serious in the northwest and may have worsened slightly in the northeast. On the other hand, China won victories against alkalinity and salinity in two coastal strips.

• The threat of acidity in subtropical and tropical areas remains, but without any one clear trend over time.

• The topsoil layer probably did not grow significantly thinner between the 1930s and the 1980s in either China or Indonesia. At first glance it would appear to have done so in both countries, but a change in the purpose and definitional practice of soil sampling explains virtually all of the evidence for thinning.

On Their Causes

• Control over pH depended on water control and, in Indonesia, on adapting liming to Indonesian conditions.

• The shifting balance between two forces that accompanied the intensification of agriculture—crop removals of soil nutrients versus the application of fertilizers—significantly shaped trends and geographic patterns in soil macronutrients (here N, P, and K).

• The main set of forces shaping nutrient trends in agricultural soils since the 1930s was *not* soil erosion.

On Changes in Cultivated Area

• In both countries, total cultivated area has risen, though not as fast as overall human population. For China the expansion of total cultivated area has been revealed only in the 1990s, after many writers had committed themselves to the view that China's agricultural land base was shrinking.

• Urbanization and other nonagricultural construction have taken only negligible shares of China's land, even if allowance is made for the fact that the lands lost to construction are of higher than average soil quality.

On the Consequences of Trends in the Land Endowment

• The stability or decline in soil organic matter and total nitrogen has had no clear effect on agricultural yields, as far as statistical explorations using data from both countries can reveal.

• The level of total potassium seems to have mattered to yields in both countries.

• Weighting each soil characteristic by its marginal impact on agricultural yields and multiplying these marginal impacts by the stock of cultivated land produces a measure of the endowment of agricultural soil capital. Changes in this endowment over time represent the net investment in productive soils. For China and in Java, the net investment in soil was probably not negative. That is, there was no net soil degradation. For the outer islands of Indonesia, the average quality of cultivated soils did decline, but the area under cultivation rose so fast that the total soil endowment rose even on the outer islands.

• In both countries, the total farmland endowments will probably not decline much in the near future, as the prevailing pessimism has feared. But in the best of worlds, the cultivated area *would* decline for China, at least the area devoted to food grains, and the expansion in Indonesia's cultivated area should taper down toward zero. Sound development policies would bring these desirable, but seldom desired, slowdowns in cultivation while maximizing the supply of food to both countries.

These initial conclusions, and their apparent implications for the relationship of economic development to environmental quality, can emerge only after a sober appraisal of the limitations of old evidence and both the limitations and the potential of the new evidence presented here. Judging the quality of the evidence is the task of the next two chapters.

2

Previous Evidence

The ratio of conjecture to hard fact is disturbingly high when it comes to modern trends in assessing the quality of the land we work. The "well-known facts" are often not facts at all. The complacent view, that land loss is no major problem, has little basis beyond the casual observation that agricultural productivity has risen. The opposite view, that the alarm must be sounded, has drawn on a wider array of suggestive evidence. Yet it too is shaky. To see why, we need to review how indirect and circumstantial are the main exhibits in the case for alarm.[1]

This chapter reviews and critiques the evidence offered to date on land loss and land quality. Its tone is negative and cautionary on account of the serious flaws in the nonhistorical contemporary literature concerning major trends in the supply and quality of cultivable land. We begin with the kind of evidence that is the grandest, the least reliable, the least historical, and the most influential today. Much of this influential literature should be discarded. By the end of the chapter we will have moved to types of previous evidence that are more modest and more reliable in the limited things they say about long-run soil trends.

Surveys of Degraded Land Areas: The Still-Life as a Documentary Film

What looks at first like a summary of what we need to know about soil degradation is already available for the whole world, and separately in more detail for China and Indonesia. For the world, for China, and for Indonesia, we have apparent censuses of degraded lands.

The single document that shapes most of the current global conjectures is the *World Map of the Status of Human-Induced Soil Degradation*, compiled in the late 1980s for the United Nations Environment Program and other agencies. The map is part of the ongoing Global Assessment of Soil Degradation (GLASOD) (Oldeman, van Engelen, and Pulles, 1990; Oldeman, Hakkeling, and Sombroek, 1991). With its accompanying explanatory handbook, it offers a rich harvest. It shows more than just the soil degradation trends and where they are occurring around the globe: It names the processes of degradation, with heavy emphasis on wind and water erosion, and it also tells us their cause, as evident in the words "human-induced" in the title. The map and the tables based on it even tell us when these trends have been occurring, namely since 1945.[2]

Tables 2.1 and 2.2 offer a quantitative manifestation of the GLASOD map. The amount of degraded lands seems extensive, comprising about 23 hectares of human-degraded land for each 100 hectares still tilled or pastured or forested as of 1991. The World Resources Institute's digitization of the GLASOD map is quite explicit about both the timing and the causes of the soil degradation. In table 2.1, the World Resources Institute explicitly dates the soil degradation as "1945 to late 1980s." And table 2.2 breaks the alleged human causes of the degradation down into further detail. Thus we are shown both the form and the human cause of all the world's soil degradation.

The institute's dating of the human-induced degradation should set off a warning bell, even if the mentions of causes and places have not. How could either the GLASOD team or WRI know how good the soil was at any particular point in the past? What do we know about the state of the world's soil in 1945, other than the fact that a war ended that year? No global soil survey was conducted in or near 1945. In fact, there was never a global soil survey worth the name until the GLASOD mapping effort in the late 1980s. How can we know whether a change has taken place, without at least two dates between which to measure the change?

Another wave of doubt surrounds the idea that the changes were "human -induced." If we don't know with any real degree of certainty when the soil changed, or even whether it changed, how can we know why it changed? The preannounced purpose of the project mixed prejudgments of causes and remedies into the scientific inquiry itself: "Strengthening the awareness of policy-makers and decision-makers of the dangers resulting from inappropriate land and soil management, and leading to a basis for the establishment of priorities for action pro-grammes" (Oldeman, Hakkeling, and Sombroek, 1991, Explanatory Note, p. 2).

It takes great effort for careful soil scientists to sort out human from natural contributions to erosion and other soil changes in their detailed studies of each locality. To globalize the effort, the GLASOD team used the best resources available on a global scale. That is, they sent data requests to experts around the world, asking for a response within nine months. They asked each to consider their local terrain, climate, current soil state, and other conditions, and to produce from these a judgment of how fast the different kinds of soil degradation might be occurring.[3]

This survey procedure yields not data on how the soil has changed over time, or why, but rather a set of predictions on how much degradation might occur for a particular kind of terrain, climate, and so forth, given the available side evidence from field experiments from around the world. The predictions could easily miss the mark. Humans are constantly affecting the soil in complex ways. And as this book emphasizes, humans react to the condition of the soil itself in ways that are meant to change the soil. None of the available predictions and simulations captures this complexity of human responses. Furthermore, the definitions of areas provided to the experts are not in keeping with the "human-induced" packaging of the whole project. At the global project level, the authors of the GLASOD map meant the phrase "human-induced degraded areas" to contain a redundancy. By their definition, there are no degraded areas that are not human induced. Unvegetated wastelands fall outside the "degraded" category. Yet the instructions to the experts defined water and wind erosion as *any* water and wind erosion, without reference to human action. This may be one reason why erosion emerged in the project as the largest cause of land degradation.

The most ambitious attempt to date to map global soil trends, then, seems to have three main drawbacks:

1. It tries to measure changes over time in the absence of data over time, attempting to run a whole documentary film from a single picture.

2. The picture used is closer to an artistic still-life painting than to a real-world snapshot, in that it is a set of predictions and not a census of soil conditions on nonexperimental farms.

Table 2.1
The GLASOD-WRI global inventory of forms of human-induced soil degradation, 1945 to late 1980s (in millions of hectares)

	World	Africa	North and Central America	South America	Asia	Europe	Oceania
Total degraded area	1,964.4	494.2	158.1	243.4	747.1	218.9	102.9
Water erosion							
Total	1,093.7	227.4	106.1	123.2	439.6	114.5	82.8
Topsoil loss	920.3	204.9	80.9	95.1	365.2	92.8	81.7
Terrain deformation	173.3	22.5	25.2	28.1	74.4	21.8	1.1
Wind erosion							
Total	548.3	186.5	39.2	41.9	222.2	42.2	16.4
Topsoil loss	454.2	170.7	37.5	22.7	165.8	42.2	16.4
Terrain deformation	82.5	14.3	1.7	18.4	47.5	0.0	0.0
Overblowing	11.6	1.5	0.0	0.8	8.9	0.0	0.0
Chemical degradation							
Total	239.1	61.5	7.0	70.3	73.2	25.8	1.3
Nutrient loss	135.3	45.1	4.2	68.2	14.6	3.2	0.4
Salinization	76.3	14.8	2.3	2.1	52.7	3.8	0.9
Pollution	21.8	0.2	0.4	0.0	1.8	18.6	0.0
Acidification	5.7	1.4	0.1	0.0	4.1	0.2	0.0
Physical degradation							
Total	83.3	18.7	5.9	7.9	12.1	36.4	2.3
Compaction	68.2	18.2	1.0	4.0	9.8	33.0	2.3
Waterlogging	10.5	0.5	4.9	3.9	0.4	0.8	0.0
Subsidence of organic soils	4.6	0.0	0.0	0.0	1.9	2.6	0.0

Table 2.1
(continued)

	World	Africa	North and Central America	South America	Asia	Europe	Oceania
Undegraded areas							
Permanent agriculture and stabilized terrain	6,092.0	1,305.0	886.0	1,143.0	1,692.0	624.0	441.0
Natural area	3,486.0	435.0	1,019.0	354.0	1,329.0	106.0	243.0
Nonvegetated land	1,469.0	732.0	128.0	28.0	485.0	1.0	95.0
Total degraded and undegraded lands	13,011.4	2,966.2	2,191.1	1,768.4	4,253.0	949.9	881.9

Sources: Map and Explanatory Note by Oldeman et al. (1991), as digitized into land areas by World Resources Institute (1992), tables 19.3 and 19.4, pp. 290–95.

Notes: The total land area covered is all areas between 72 degrees north and 57 degrees south. The division between Asia and Europe is as described by the World Resources Institute: "U.S.S.R. east [west?] of the Ural mountains is included under Europe, and U.S.S.R. west [east?] of the Ural Mountains is included under Asia." Details may not add up to totals due to rounding. In the original sources, the total degraded areas are broken down into areas of light, moderate, heavy, and extreme degradation.

Table 2.2
The GLASOD-WRI global causes of human-induced soil degradation (in millions of hectares)

	World	Africa	North and Central America	South America	Asia	Europe	Oceania
Total degraded area	1,964	494	158	243	748	219	103
Vegetation removal	579	67	18	100	298	84	12
Overexploitation	133	63	11	12	46	1	0
Overgrazing	679	243	38	68	197	50	83
Agricultural activities	552	121	91	64	204	64	8
Industrial and biotechnical	23	0	0	0	1	21	0

Sources: See sources for table 2.1.
Notes: See notes to table 2.1.

3. It proceeds from its own preconception that all land degradation is human induced.

Indonesia and China have tried to conduct their own surveys classifying all lands into degraded and undegraded. Table 2.3 reproduces a sketch of the land areas of Indonesia in 1987. According to the estimates in this table, Java's land degradation is not outstanding by the standards of the whole archipelago. Kalimantan and Irian Jaya were relatively undegraded in 1987, though they have been cleared and degraded considerably since then. Other areas, like the remote sides of Sumatra and Sulawesi, were apparently heavily degraded even as of 1987. Yet for Indonesia's 1987 survey the same questions arise as for the global survey: How do we know that humans are responsible for the extent of land degradation estimated here? What role did cultivation of parts of Indonesia play? And when did the degradation happen?

The same questions arise regarding the now-annual censuses of degraded lands in China. The estimates from China's Ministry of Water Resources and Electric Power, illustrated in table 2.4 and figure 2.1, might give the impression that we now have a quantitative history showing the amounts of land degraded annually since 1979. Erosion in particular marches upward, to judge either from table 2.4's aggregate for eighteen provinces and three municipalities or from the slightly larger estimates for the whole nation shown in Huang and Rozelle 1995. This slow upward march is also suggested in figure 2.1's curve for all twenty-one local governments. For one who believes that humans are causing more and more erosion, figure 2.1 might seem to offer a confirmation. The slow but unmistakable rise of reported erosion areas for all twenty-one governments seems to confirm that humans continue to erode the land, with the total erosion areas already greater than the total cultivated areas.

Yet the appearance deceives. On closer examination, what looks like a motion picture of erosion in action is really just a single snapshot, with some confusion added. To see why the ministry's time series does not represent a true erosion dynamic, consider the behavior of the series separately reported by provincial and municipal government agencies. Let us take the case of five western provinces, where we know erosion has been truly widespread since even before human settlement. Figure 2.1

Table 2.3
Indonesia: Degraded lands, April 1987 (in thousands of hectares)

Province	Total areas			Degraded land					
	Province	Forest	Agricultural land	In forest	Percentage of forest degraded	Outside forest	Percentage of agricultural land degraded	Total	Percentage of province degraded
Aceh	5,539	3,282	2,144	54	1.6	355	16.5	408	7.4
North Sumatra	7,079	3,526	3,876	200	5.7	708	18.3	907	12.8
West Sumatra	4,978	2,943	1,644	6	0.2	85	5.2	91	1.8
Riau	9,456	6,546	3,213	4	0.1	234	7.3	238	2.5
Jambi	4,492	2,614	2,107	5	0.2	41	1.9	46	1.0
South Sumatra	10,369	4,028	5,807	108	2.7	461	7.9	569	5.5
Benkulu	2,117	991	862	327	33.0	225	26.1	552	26.1
Lampung	3,331	1,244	2,129	206	16.6	46	2.2	252	7.6
Sumatra	47,361	25,174	21,783	910	3.6	2,155	9.9	3,065	6.5
Jakarta	59	1	45	0	0.0	0	0.0	0	0.0
West Java	4,630	974	3,414	63	6.4	242	7.1	305	6.6
Central Java	3,421	674	2,546	0	0.0	181	7.1	181	5.3
Yogyakarta	317	16	278	1	6.3	8	2.9	8	2.6
East Java	4,792	1,348	3,179	17	1.3	329	10.4	346	7.2
Java	13,219	3,013	9,462	81	2.7	760	8.0	840	6.4
Bali	556	126	407	9	7.1	45	11.1	54	9.7
West Nusa Tenggara	2,018	1,064	905	63	5.9	197	21.8	260	12.9
East Nusa Tenggara	4,788	1,487	2,869	885	59.5	889	31.0	1,774	37.1

Table 2.3
(continued)

Province	Total areas		Agricultural land	Degraded land					
	Province	Forest		In forest	Percentage of forest degraded	Outside forest	Percentage of agricultural land degraded	Total	Percentage of province degraded
East Timor	1,487	690	n.a.	73	10.6	36	n.a.	109	7.3
Bali and Nusa Tenggara	8,849	3,367	4,181	1,030	30.6	1,166	27.9	2,197	24.8
West Kalimantan	14,676	7,695	5,179	720	9.4	184	3.5	904	6.2
Central Kalimantan	15,260	10,997	1,758	n.a	n.a.	n.a.	n.a.	n.a.	n.a.
South Kalimantan	3,766	2,030	1,717	155	7.7	95	5.5	250	6.6
East Kalimantan	20,244	15,952	1,908	n.a	n.a.	n.a.	n.a.	n.a.	n.a.
Kalimantan	53,946	36,674	10,562	875	2.4	279	2.6	1,154	2.1
North Sulawesi	1,902	1,584	967	72	4.6	234	24.2	306	16.1
Central Sulawesi	6,973	4,166	1,823	249	6.0	242	13.3	491	7.0
South Sulawesi	7,278	3,352	3,363	224	6.7	267	7.9	491	6.7
Southeast Sulawesi	2,769	2,190	1,703	397	18.1	139	8.1	536	19.4
Sulawesi	18,922	11,292	7,855	942	8.3	881	11.2	1,823	9.6
Maluku	7,451	5,097	2,828	281	5.5	307	10.9	588	7.9
Irian Jaya	42,198	28,816	11,641	n.a	n.a.	n.a.	n.a.	n.a.	n.a.
Maluku and Irian Jaya	49,649	33,913	14,469	281	0.8	307	2.1	588	1.2
Total Indonesia	191,944	113,433	68,312	4,120	3.6	5,548	8.1	9,667	5.0

Source: Soemarwoto 1991, pp. 214–15, citing data from Buro Pusat Statistik.

Table 2.4
Reported degraded land areas in China, 1979–1995 (in thousands of hectares)

Year	Salinity land	Erosion land	Flood affected	Drought affected
1979	6,189	110,001	1,655	7,938
1980	6,017	109,846	4,479	9,471
1981	6,166	110,116	3,459	9,984
1982	6,167	110,402	3,941	7,779
1983	6,228	111,040	4,623	6,138
1984	6,181	111,742	4,467	5,380
1985	6,206	118,823	4,676	8,275
1986	6,174	120,910	2,639	12,149
1987	6,163	121,595	3,296	10,142
1988	6,247	123,286	4,914	12,918
1989	6,063	118,729	4,575	9,487
1990	6,070	114,837	5,045	6,117
1991	6,157	119,977	13,375	8,061
1992	6,178	125,071	3,128	14,806
1993	6,182	125,466	6,267	7,215
1994	6,173	125,466	5,820	14,616
1995	6,181	125,466	4,609	8,341

Sources: Ministry of Water Resources and Electrical Power's estimates of land areas subject to these four types of degradation, as cited and described in Huang and Rozelle 1995. I am indebted to Scott Rozelle for providing an updated data file of the Ministry's estimates by province.
Notes: Totals include the municipalities of Beijing, Tianjin, and Shanghai plus eighteen provinces. The following provinces are not included: Nei Mongol, Jilin, Liaoning, Jiangxi, Guangdong, Guangxi, Qinghai, and Tibet.

illustrates the behavior of the erosion area numbers for five provinces reporting erosion areas that far exceed their reported areas under cultivation. Each erosion area is divided by the fixed cultivated area of 1983, as a way of standardizing relative erosions. Even a glance at the movement of erosion area for these five provinces shows that something strange and artificial is at work. Gansu province, stretching from the loess plateau northwest into the desert, had zero erosion in many years, intermixed with years in which reported erosion areas either fell or jumped. They jumped in 1985, and then remained steady for another three years. In 1989 and 1990, they suddenly fell by more than the total cultivated area

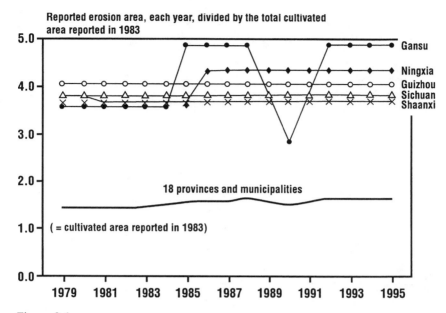

Figure 2.1
Reported erosion areas in China, 1979–1995, each relative to the cultivated area reported in 1983
Source: See table 2.4
Note: The cultivated areas used to define 1.0 on the vertical axis are (in thousands of hectares): 76,589 for all eighteen provinces and three municipalities; 1,896 for Guizhou; 6,525 for Sichuan; 3,749 for Shaanxi; 838 for Ningxia; and 3,552 for Gansu.

of Gansu (that is they fell by more than 1.0 in figure 2.1). Then in 1991 and 1992, they jumped by the same amounts, returning to the old high level and staying fixed there. Can one believe that this represents the true behavior of Gansu's eroded areas? More likely, it represents the provincial authorities' struggle to define something hard to define: a strict dividing line between erosion areas and non–erosion areas. Apparently they changed their minds a few times, perhaps in response to changing instructions from the central government. It is hard to imagine that a true erosion process would behave like the erosion areas reported from Gansu.

The same problem shows up in the numbers for the other four highly eroded provinces. Their officials reported zero erosion in most years, punctuated by the occasional discrete change, like the jump in reported erosion in Ningxia in 1986. Again, this looks like a reporting problem, not the steady creep of wind and water erosion.

In fact the erosion area numbers for all provinces behave strangely. In 54 percent of the cases, the year-to-year change at the province level is reported as zero. When the erosion area does increase, it takes an implausibly large jump. Simply summing only one year's increase (the largest year-to-year increase out of seventeen years) for each province or city accounts for fully 60 percent of the whole 1979–1995 net increase in erosion area for these eighteen provinces plus three cities. True wind and water erosion does not behave this way.

The erosion area figures seem to have been provincial officials' best attempt to carry out a nearly impossible task. In general, making land degradation a binary yes-no variable for each land area is very difficult. Trying to do so threatens to make the share of lands degraded jump or plummet with slight movements in the definitional threshold of "eroded." The problem is a familiar one in social science. The share of the population that is "urban," for example, can rise or fall sharply with small changes in the definition of what is a city and what is not. Land area in forests is very sensitive to the degree of canopy cover that defines a forest, and shares of women with jobs are very sensitive to just how many hours of part-time paid work constitute having a job. One should therefore be hesitant to read any trend into the ministry's figures on degraded land areas.[4] Instead, the figures represent a single snapshot for each province, with a slightly blurry focus: Erosion was a fixed amount for each province, plus or minus an arbitrary change when officials changed their definitions of "erosion."

Guesswork on Erosion

The larger literature on erosion trends has not depended on government time series on degraded areas, like those from China's Ministry of Water Resources and Electrical Power. Rather, it has depended on even weaker data.

Global Soil Loss Rates

To show how quickly land is being lost, most writers start with tonnages of erosion losses or with measures of the land areas lost. The erosion estimates often feature data on river turbidity and siltation. Judson (1968) estimated that the world's rivers carry 24 billion tons of sediment to the ocean each year, compared with only a geological rate of 9 billion tons before the introduction of agriculture, grazing, and other human activities.

Many such sketchy measures of world soil loss rates have been presented with confidence (and occasionally with disclaimers). In retrospect, they should have shown less confidence. For example, the World Resources Institute and the United Nations warn that "for those who contend that land productivity around the world has deteriorated, anecdotal evidence is compelling. But there are few hard data to buttress the case. And despite the impetus of [the United Nations Conference on Desertification] little progress has been made toward an accurate worldwide assessment of how much degraded land exists" (World Resources Institute 1988, p. 13). Despite their having sounded this warning, they proceed immediately to estimate the rate of soil erosion for twenty-four developing and three developed counties for 1970–1986 on the basis of indirect guesswork and find that "clearly, far more topsoil is lost from cropland each year than is replaced naturally. An estimated six to seven million hectares of agricultural land are now rendered unproductive each year because of erosion—more than twice the rate in the past three centuries."[5] Several other studies offer similar estimations for tropical countries (Lal 1990, p. 15). And Brown and Wolf (1984) warn that "excessive erosion" took 4,300 million tons a year from 245 million crop acres in China, and 4,700 million tons from 346 million acres in India. In all cases, the figures seem to warrant alarm.

More extreme is the U.S. Department of Agriculture's (Dregne 1982, p. 5; Sutton 1989, p. 36) matrix of six continents' soil degradation rates. Although the "rate" is not defined clearly in units of time, the USDA study assigns productivity costs to it, implying that soil degradation has cut farm output on six continents by something between 8 and 30 percent.[6]

These figures seem to come from two kinds of sources: (1) purely experimental data on how much soil could be lost if certain conditions held in the real world, like the experimental data on which much of the GLASOD map of human-induced soil degradation was based, and (2) river siltation studies.

The leading attempt to calculate the costs of soil erosion in Java offers an example of how experimental data can be misapplied. Robert Repetto and his colleagues based their study (Repetto et al. 1989) primarily on an unpublished World Bank study by William Magrath and Peter Arens (1987) that acknowledged it had no good basis for its calculations, made them anyway, and cited no published works. The closest thing in the World Bank study to a measure of soil degradation is based on experimental work, not real-world observations from Java.

Magrath and Arens's definition of soil degradation involves soil chemistry conditions, and the authors prepare the reader for a modest exercise that must use indirect methods: "Soil degradation is a gradual process that occurs as soil depth declines by erosion leaving progressively less topsoil and lower nutrient concentrations. Data monitoring this process on aggregate levels in Indonesia are not available. In this section alternative indicators of degradation are reviewed along with estimates of their quantitative significance on Java" (Magrath and Arens 1987, p. 2).

Magrath and Arens's method starts with a table of the land areas of four regions of Java, categorized by twenty-five soil types and eleven erosivity classes, from a 1959 Food and Agriculture Organization (FAO) paper. Within each of the four regions, digitized maps estimate the area in each combination of 25 soil classes × 11 erosivity classes × 4 land uses. Apparently experimental data and other assumptions predict soil loss rates for each combination, in metric tons per year.

Their next task is to estimate the productivity consequences of erosion. Magrath and Arens are sober about this next step as well (1987, p. 10): "While it is widely accepted that erosion lowers agricultural productivity, there is little agreement on exactly how productivity is related to erosion or on the quantitative impact of erosion on yields." What to do? Their table 11 (p. 19) cites three local studies from Java giving "annual" (cumulating?) percentage changes in yields from some kind of erosion that may or may not be the same as what their predictions just calculated. It is also courageous for them to project three local studies onto their 25 × 11 × 4 = 1,100-cell grid. Worse, they list only one of the three sources mentioned in that key table: a 1984 project report not available to the public.

This Magrath-Arens experiment-based exercise is the basis for the cost estimates of Repetto et al. (1989), which Barbier and Bishop (1995) also reproduced, as illustrated in table 1.1.

River Silt

The second basis for estimates of soil losses is the volume of soil conveyed through rivers and other watercourses per day or per year. River siltation rates, however, for all their popularity in the literature, are virtually unusable as evidence of human-induced erosion. They mislead by failing to alert readers that the soil losses they measure (1) come from entire watersheds, not just from the small parts of them worked by humans, and (2) reflect gross soil losses, often from unknown places, rather than the net losses in soil depth or in soil nutrients at any one place. The distinction is crucial where tillage and fertilization are separate influences on soil depth and quality.

Some examples from China and Indonesia show the potential for distortion. In Wen Dazhong's stimulating study of erosion in China (1993, pp. 65–71), the total areas losing soil to rivers and the atmosphere are large multiples of the areas actually cultivated there. In Wen's study, cultivated lands accounted for only 30 percent of the soil loss on the loess plateau and only 9 percent of the soil loss in the southern erosion region.[7] The same is true of Indonesia's erosion rate aggregate studies: They too are based on the river-borne silt from entire watersheds, with most of the land area uncultivated.[8]

A similar casual treatment of geography and history pervades the literature on desertification. For China, the centuries-old concern about advancing deserts can be illustrated even by going no farther back than the 1930s, which is the starting point for this book's historical coverage. One of the pioneers of the data used here, James Thorp, described in 1939 the state of the war in the ecological battle zone of the loess plateau that extends from Qinghai province in the west through parts of Gansu, Ningxia, and Shaanxi into Nei Mongol (Inner Mongolia):[9]

In geologically recent times and with great acceleration, since the advent of agriculture a few thousand years ago [the loess plateau's] loess deposits have been severely eroded, so that now the thicker of them are cut by innumerable vertical-walled gorges and gullies which, in spite of almost super-human efforts on the part of the inhabitants, are rapidly reducing the loessial plateau to a region of 'badlands.' Only the divide region of north Shensi [Shaanxi] and some of the terrace remnants in the Wei Basin retain their former smooth topography.

The loessial plateau grades off northward into the rocky and sandy Ordos Desert which doubtless was the source of much of the dust which now makes up the loess deposits. At the present time the sands of this desert region are gradually creeping southeastward with the prevailing northwest winds of the winter months, and in some places they are covering the habitations and farm lands of the people. There is much evidence to indicate that some of these sand dunes have been caused by the plowing of sandy grasslands by Chinese farmers who encroached on the grazing lands of the nomads. Grass roots are able to prevent the development of sand dunes from sandy semidesert soils, but ordinary agricultural crops cannot do this.[9]

What looked dangerous in the 1930s has continued to look dangerous in the 1980s and 1990s. Most accounts of the soil issue at the global level (Eckholm 1976; United Nations Conference on Desertification 1977; Smil 1987; Forestier 1989; Dregne 1992, pp. 9–10) have featured desert threat in the loess plateau and farther west and north. So has the Western press, as dramatized by this account in the *Washington Post* in November 1998:

Like a Pac-Man run amok, deserts are eating northern China. A report last year from the Forestry Ministry said China's desertification was among the most serious in the world, costing the country $6 billion a year. More than 1 million square miles, equal to nearly one-third of China's land mass, have turned from grasslands to dust over the years because of overgrazing, clear-cutting of forests and over-ambitious development policies. About 170 million people live in desertifying regions.

Each year, a wasteland about the size of Rhode Island emerges like a giant stain amid China's prairies. . . .

China's deserts have been growing for centuries. More than 2,300 years ago, the Chinese philosopher Mencius warned about overgrazing and clear-cutting. But since 1949, things have worsened at an exponential rate. The frequency of sandstorms, which are linked to creeping deserts, has increased eightfold since the 1950s. Last April 14 and 15, a mud rain, caused by a giant sandstorm, splattered most of northern China and parts of the Korean peninsula with a coat of dirt paint.

Desertification means China's yellow earth is eroding at an alarming rate, creating a potential catastrophe downstream. Each year 66 tons of dirt per acre flow from northern China toward the Pacific Ocean, 1.8 billion tons in all.

The Yellow River got its name from this eroded soil. These days, by the time the river meanders downstream to Kaifeng, in central Henan province, for example, in some places it is five stories higher than the city of 500,000 people. It flows through the city in an elevated riverbed that is blocked from flooding by massive embankments. (Pomfret 1998)

There are many things wrong with such reporting. For example, the *Washington Post* was willing to let its readers think that the desertification of fully one-third of China's land mass is both recent ("since 1949 things have worsened at an exponential rate") and due to humans ("overgrazing, clear-cutting of forests, and overambitious development policies"). Its readers are invited to imagine that there was very little Takla Makan–Gobi Desert and a well-forested loess plateau in the northwestern China of 1949. We know better, thanks to archaeological, botanical, fossil, and historical records. These arid zones were even drier than today several times in the past, including 1000–2000 B.C., again around the time of Christ, again around A.D. 1700, and again in the early nineteenth century (UN Conference on Desertification 1977, p. 124).[10]

Changes in Cultivated Area

Another indicator that is frequently cited as evidence of soil problems— but should not be—is the net change in a region or country's cultivated area.[11] We'll see in chapter 6 that in the case of China, the official figures on cultivated area used by past writers are seriously in error. More fundamentally, though, even perfect measures of cultivated area cannot reveal what is happening to the quality of the land being worked or the processes of soil change.

A decline in cultivated area does not necessarily show soil degradation or even a drop in the supply of land available to agriculture. Even improving soils can go out of cultivation if the relative prices of farm products drop or farm wage rates rise faster than farm labor productivity. Such trends are very possible, since economic progress shifts demand away from agricultural products (Engel's Law) and raises wages in response to economy-wide productivity trends. An increase in cultivated area is equally ambiguous as an indicator of soil quality.

In fact, it is surprising to see how often the changes in cultivated area are cited as evidence of trends in soil quality. Most of the citations come from pessimists, but not all: optimist Julian Simon made changes in

Table 2.5
Trends in cultivated area do not reveal trends in soil quality

If we see that	A pessimist can say this shows that	An optimist can say this shows that
Cultivated area rose	Population pressure forced people to work poor lands.	Progress reclaimed and improved new lands.
Cultivated area fell	Degraded lands had to be abandoned.	Progress pulled valuable labor and capital away from marginal lands.

cultivated area the centerpiece in his case against pessimism (Simon 1981, chap. 6). Neither side can win an argument by citing trends in cultivated area. Trying to cite them inevitably leads to a stalemate, whether the area is rising or falling, as table 2.5 shows.

To be useful, figures on land area trends must be supplemented with information on land quality and on either land price trends or its productivity, plus information on causal factors. In an ideal market economy, we could statistically analyze the rental price, or the purchase price, of land for given quality to determine why its value rises or falls. Looking at this price plus the change in area can help us sort out the shifts in land demand and land supply—including soil quality—that caused both the price change and the land area change. For China and Indonesia, however, we must infer what is happening to the productivity of land without the help of good data on its rental price or purchase price. Once we have done that, in chapter 5 for China and chapter 8 for Indonesia, we can judge what forces have shifted both the value of land and its cultivated area. But using cultivated area alone will not work.

Both writers and politicians have succumbed to the temptation to use images when dramatizing land losses. A favorite image is the golf course, which replaces grain land with a spacious playground for the rich. *China Daily* has featured golf courses as a threat to China's food supply, and so have the *New York Times, Newsweek,* the *Washington Post,* and Worldwatch Institute.[12] In 1995 *Newsweek* warned that "by the end of the century, China could have 100 golf courses, up from a dozen today." Worldwatch Institute thinks that the golf threat is even greater:

One of the most land-intensive leisure-time activities is golfing. As incomes rise in Asia, the number of golfers is rising at an extraordinary rate. This has led to the construction of thousands of golf courses in Japan, the Philippines, Indonesia, Viet Nam, Malaysia, and China. Sensing the threat this poses, Viet Nam has banned the construction of golf courses on rice-land. Similarly, Guangdong Province, the southern coastal province of China, which has been the fastest-growing province in China, also had to ban golf course construction to protect its cropland. (Brown et al. 1998, pp. 80–81)

How many golf courses would it take to starve China? To grasp the magnitudes behind the imagery, suppose first that *Newsweek*'s fears are confirmed, and China has developed 100 golf courses by the year 2000. Suppose also that those courses are built on the land-intensive American standard of 42 hectares per golf facility.[13] Under these assumptions, the 100 courses could deprive China's agriculture of 4,200 hectares, or less than 0.003 percent of China's cultivated area in 1995. To take even this tiny loss seriously, one would have to imagine that China's golf course employees do not receive any income at all from the golf course and therefore cannot buy more food. To go beyond *Newsweek*'s fear of 100 courses in China to the more extravagant Worldwatch fear of "thousands" of golf courses would allow one to imagine golf courses taking up as much as 0.03 percent or even 0.05 percent of China's agricultural land. In an unrealistic extreme, if China were just like rich and land-abundant America and has 12,800 golf facilities taking up 532,700 hectares, golf would take up less than 0.4 percent of China's cultivated area. But China will never develop anything like the amount of golf course area the Americans have, and *Newsweek*'s tiny 100 golf courses consuming 0.003 percent of China's cultivated area is much closer to the mark. Any local ban no golf courses in the name of the food supply is therefore political symbolism without substance.

Lessons from Experiments

Classic Long-Term Experimental Sites
Impressive pioneering efforts in agronomy and soil science have in fact left us with a long historical record of soil conditions and agricultural yields—on experimental plots. Starting with Rothamsted, England, in 1843, this grand tradition was continued by the Morrow plots at the

University of Illinois in 1876, and by several other long-term experimental plots since.[14]

The classic experiments confirm the basic difference in change patterns between forest and cleared land. Clearing a forest for field crops cuts organic carbon and nitrogen and often raises pH toward neutrality. The "wilderness plots" at Rothamsted, which gained organic matter for decades after they were first left alone in the late nineteenth century, offered a sort of reverse confirmation of this pattern of organic matter loss. Such loss also shows up unmistakably in recent experience, especially in the tropics (e.g., Lal and Pierce 1991, p. 139). We shall revisit this land-clearing effect in our exploration of Indonesia, where it plays an important role in interpreting soil trends.

The long-term sites have served their original purpose by revealing the dynamic effects of varying fertilizer regimes and crop rotations on yields and on soil chemistry. Rothamsted, Morrow, and other sites confirmed the sustainability of yields in temperate-zone monocultures based on inorganic fertilizers. Reapplying inorganic fertilizers after they had been withheld from a monocultural plot restored yields "immediately and dramatically." While the best yields require both soil treatments and crop rotations, the soil treatments were found to matter more than the cropping system. An important by-product of vigorous crop growth was protection against soil erosion (Jenkinson 1991, p. 3; Mitchell et al. 1991, pp. 24–27).

The long-term experimental sites have proven to have their limitations, however. Aside from being expensive and being less valuable until the experiments were randomized and replicated, the long-term plots had a familiar problem: keeping relevant in a changing human world. By sticking to fixed regimes over the decades, the experiments became obsolescent as humans changed their agriculture in unforeseeable ways: "The dilemma is obvious: leave the experiment unchanged and allow it to become irrelevant to agriculture or change the experiment and lose continuity with the earlier years. The Rothamsted solution is to do both: to modify the experiment in the light of contemporary agriculture, but *slowly,* after the changes have become well established and, as the same time, to maintain some cropping sequences, fertilizer treatments etc. as near as possible to the original design" (Jenkinson 1991, p. 9). The problem and the solution are thus analogous to the economist's problem of adjusting a historical cost-of-living index as new consumer products replace obsolete ones: Find the optimal gradual way to splice different comparisons together over the years.

One way to keep a longitudinal study up to date is to shorten the experiment's time span. By looking only a dozen or so years into the recent past, a research team can collect a wider variety of data on the experimental plot and even extend the study to cover the separate plots of autonomous farmers. Kenneth Cassman and his coauthors (1995, 1996) have done just that for lowland flooded rice cultivation in the Philippines. Working under the auspices of the International Rice Research Institute and affiliated research units, they monitored changes in soil and rice yields on both experimental plots and nearby farmers' plots. The results are generally cautionary and remain relevant under current practice. Adding nitrogen fertilizers has its limits in the lowland rice zone, and yields have stagnated or declined. Over the years repeated applications have little effect on total-soil organic matter or nitrogen, and even less on the available nitrogen, in this anaerobic setting. Although repeated application can uphold yields within the year of application, getting the application regime right proves complex. The linkages involving soil organic matter and nitrogen thus appear quite different from those in aerobic cultivation, posing a challenge to the goal of continuing to raise rice yields.

In the end, since the value of any longitudinal experiments depends on how well they imitate the changing choices farmers face, we still need to confront the importance of establishing a truly nonexperimental historical database, one that allows us to follow how real-world agricultural practices relate to the soil. This nonexperimental database is featured in the chapters that follow.

Topsoil Depth Experiments

More popular—and more dangerous—than longitudinal agronomic experiments is the use of topsoil depth studies to quantify the effects of topsoil loss on agricultural yields. Like the classic long-term agronomic experiments, the topsoil depth studies have the problem of staying relevant to actual farming practice. But unlike agronomy's use of the classic experiments, the debate on soil trends often misuses topsoil depth experiments in ways that flunk the relevance test.

Topsoil depth experiments come in two basic varieties. Some vary topsoil depth spatially, holding site attributes as constant as possible. Others remove topsoil from the same site, and then replow and recultivate and measure the effect on yields. Surveys of this literature[15] have found that there is a yield effect from extra topsoil up to a depth of 40 centimeters, beyond which the yield effect is negligible. The experiments come predominantly from the United States.

The topsoil depth experiments are largely irrelevant, however, to the debate over soil trends in the real world. Whether they vary topsoil depth spatially by comparing different sites or use bulldozers to scrape off a top layer, they fail to imitate what actual farm populations experience with or without topsoil loss. Both kinds of experiments miss a key principle: economic choice. The spatial variation across sites tries to hold too many things equal, not letting farmers adjust their inputs and soil management to the fact of differences in soil depth. The more numerous topsoil removal experiments have what we could call a bulldozer bias. On balance, one should suspect that the bulldozer experiments overstate actual yield losses from erosion. The principle of economic choice says that postbulldozer yields, with no other modifications than the experiments applied, seriously underestimate what rational farmers would have achieved under the same conditions to improve the soil. Given that they would have had longer to react and adjust to gross topsoil losses than the experimental cultivators of freshly bulldozed soil had, farmers would have maintained its productivity better.

Studies that dramatize the costs of erosion also misuse topsoil depth experiments through geographical mismatching. The depth experiments come from the United States and other prime agricultural areas. Measures of the yield effect of topsoil depth from these areas tend to give higher losses, in tons of grain per ton of topsoil, than would be appropriate to the areas where the erosion is more serious. Thus in China the danger is that in estimating economic damage, tons of soil lost from the loess plateau will be multiplied by tons of grain per ton of soil not from the loess plateau but from fertile Jiangsu or even from the United States, thus overstating the cost of erosion. The overstatement is even greater, of course, when the tons of soil loss from the loess plateau refer to the entire region, not to the cultivated areas.

Soil Conservation Studies

On a more positive note, the final kind of current evidence noted here often estimates both the costs of soil degradation and the benefits of specific soil conservation practices using extensive information for a particular locality. Good examples for our two countries are studies of the importance of soil conservation practices in southern China (Parham, Durana, and Hess, 1993; Wang and Gong 1998), Gansu province (McLaughlin 1993), and upland Java (McCauley 1985; Carson 1989; Barbier 1990; Nibberling 1991).

Such studies are superior to most of those in the soil trends literature because they develop specific costs and benefits in some detail. They are based, however, on data from just a few small localities and on experimental data rather than a history of average practice. As is noted in the postscript to this book, separate tasks call for separate kinds of research. Detailed local studies like these are essential to appraising the returns from specific soil improvements or the ending of specific soil degradation. To judge the trends in average human soil management, however, as the larger debate has tried to do, one needs a broader historical database, one as broad as the assertions being judged.

Hints from Good Eclectic Histories

To complete this survey from the grandest and worst evidence to modest solid evidence, let us close with a look at some of the more truly historical studies that have been conducted.[16]

Several studies have made an effort to construct a historical record of soil trends in a particular area or locality based on an eclectic array of indirect clues. Some are studies specifically about the soil of an area whereas others are not, and some focus on China or Indonesia whereas others do not.

The best direct predecessor of the present book is an in-depth local study of the Machakos District in Kenya, just east of Nairobi, conducted for the World Bank by Mary Tiffen, Michael Mortimore, and Francis

Gichuki (1994). One of the ways in which their study stands out is that it actually compares soil samples separated by more than a decade, in this case samples from comparable sites in 1977 and 1990.[17] Its discussion of the problem of determining truly "comparable" places at different times anticipates an extended discussion in chapters 3, 4, and 7 of this book. Tiffen, Mortimore, and Gichuki also made effective use of an eclectic database, including a series of photographs of the same slopes and valleys in 1937 and 1991. Even their conclusion, summarized in the book's title, *More People, Less Erosion,* resonates with at least one of the conclusions of this study, namely, that the study of soil degradation should shift its emphasis from erosion processes to possible overcultivation relative to inputs into the soil. This fine study is of limited use for present purposes, however, in that it mobilizes only a few direct soil data from only one place, and that place is not in China or Indonesia.

Two other non-Asian studies also showed laudable historical sleuth work in their attempt to gather every possible kind of evidence from a past that yielded no actual soil samples. Jean-Paul Harroy (1949) assembled all the scraps he could about the soil history of Sahelian Africa, the part of the world for which a soil history is most urgently needed. Stanley Trimble (1974) did the same for the southern Piedmont in the United States, an area for which he gave us every reason, short of direct soil data, for believing that deforestation and overcultivation of tobacco and other cash crops seriously depleted the soil. Though part of his evidence consisted of data on silt flows, which were criticized above, Trimble added enough other circumstantial evidence from historical geography and agricultural economics to present a prima facie case for serious economic losses from erosion.

For China, the closest thing to a long-run soil history is a persuasive article on the Tai Lake region by Erle Ellis and Wang Siming (1997). Ellis and Wang trace the region's achievement of soil nitrogen sustainability over the centuries, with a variety of adaptations to relieve the nitrogen constraint. For larger parts of China, we have the benefit of some recent major environmental histories by Pomeranz on Shandong (1993), Marks on Guangdong and Guanxi (1998), and Pomeranz on China in contrast to other regions (forthcoming, especially chapter 5). These books range over a vaster subject matter than just soil history, yet they do give reason to suspect that soil depletion was particularly serious in certain regions and certain eras.

On Indonesia, a fair picture can be woven together by combining several kinds of sources. These include Wolf Donner's (1987) provocative history of assaults on the Indonesian environment, Clifford Geertz's earlier (1963) environmental history, the agricultural histories by Anne Booth (1988) and Pierre Van der Eng (1996), the sectoral history by Peter Timmer (1987), and the time series coverage of Indonesia's fertilizer problems and policies (Adiningsih, Santoso, and Sudjadi 1990; Rosegrant and Kasryno 1990; Santoso, Adiningsih, and Heryadi 1990). The combination of these sources leads to clear conclusions (that a great deal of soil continues to make its way from the uplands to the spreading river deltas, and that agricultural productivity marches ahead) arranged around a central unanswered question: What did erosion owe to agriculture, and how much damage did it do to agriculture?

These historical studies have several properties that make us set them aside here:

• They relate mainly to earlier eras, before the great intensification of agriculture since the 1930s, which is the central concern of the current debate over soil trends.

• They are often larger environmental histories, with soil trends playing only an indirect role.

• Some are not about China or Indonesia.

• In general, they have not yet affected the recent debate over environmental trends, either because they are too recent or because they are too careful and nuanced for readers wanting strong messages.

Summary Diagnosis

The study of trends in soil quality needs the kind of quantitative history that has so enriched climatology and geology. Virtually all estimates of current soil trends, including every available map of soil degradation trends, are in fact not data. Rather they are experts' predictions, derived by combining data on slope, climate, and land use with what happens to such soils under experimental conditions. Sometimes they are refined into "expert opinion," as in the GLASOD map, but they are still not based on any observation before the mid-1980s. The only approximation to an actual time series is that series since 1975 by China's Ministry of Water Resources and Electrical Power giving areas damaged or in danger. But the areas cited in that series are not specific to farmland, and the abruptness of their occasional year-to-year jumps suggests changes in official definitions rather than changes in soil condition. We need to find new historical data, data that directly measure the quality of agricultural soils in past and present.

3
Beginning a New Soil History

Building a new basis for analyzing data on soil trends must start from a firm foundation. This chapter lays that foundation in some detail, to enable later chapters to present a clearer view of the broader history.

Global Data from the Twenty-First Century

By the middle of the twenty-first century, many of the difficulties of finding and using historical soil data should be alleviated. The kind of data assembled here for two countries will be available in ever greater abundance for at least a dozen countries around the globe, including thousands of soil profiles for each of these countries dating back to the 1980s.

That data collection is already in motion on two fronts. First, the "soil reference series" already being disseminated by the International Soil Reference and Information Centre (ISRIC) with U.N. support gives modal "soil reference profiles" for a large number of countries. These have helped to inform revisions of ISRIC's *World Map of the Status of Human-Induced Soil Degradation*. By the middle of the twenty-first century, chapter 2's warnings about the ISRIC map should become obsolete, as decades of accumulation of soil profiles will have made a multicountry soil history possible.

On a second front, in all likelihood, the new multicountry perspective made possible by accumulating data in the decades ahead will rest on richer data sets from a small number of countries, with modest numbers of summary "soil reference profiles" made available for others.[1] The research project behind this book has demonstrated that judging trends requires large numbers of soil samples. Even if each representative profile is actually a composite of many soil samples, the heterogeneity of soil still requires at least 1,000 observations to capture changes in a large region or country between two decades. It seems likely that data at the requisite level of richness required to follow a half century of change will be available for only a few countries—again, perhaps a dozen or more—having the strongest commitments to soil surveying. That set of nations will surely include China and Indonesia. If so, it will be a continuing advantage to be able to extend an abundant soil survey record for these two countries spanning from the twenty-first century back to the 1930s.

This chapter introduces the historical data sets for China and Indonesia, with suggestions and cautions about their use. Having a sound basis for judging both the data accuracy and the reliability of the average trends will make it possible to say for the first time how and where the average soil characteristics for these two regions have changed. The data will reveal plausible patterns in soil chemistry, some expected and some not. Although the data sets do not allow us to identify local changes at the level of the county or village, they do suffice to resolve some of those larger debates about soil trends over whole regions.

Mining the Twentieth-Century Record

Thanks to some pioneering efforts in soil surveying, we can learn much about the physical and chemical state of soils in China and Indonesia since the 1930s.[2]

China
China's long tradition of soil science dates back at least 2,000 years to the Guanzi descriptions of water table depths and soil fertility classes in what is now north central and east central China. For the twentieth century, the soil data have come in three waves.[3] For the period 1932–1944 (hereafter referred to as "the 1930s"), Sino-American teams based in Nanjing and in wartime Chongqing published, for the National Geological Survey of China, several hundred soil profiles from all the parts of China they could survey in such turbulent times.[4] As can be seen from table 3.1's data inventory and from the illustrative reporting of pH data points in

figure 3.1a, the geographical coverage of the 1930s wave was by necessity uneven. The northeast provinces occupied by Japan could not be covered, except for a few profiles from Heilongjiang, nor could areas that were militarily insecure, such as the warlord-dominated parts of Guangdong, where the research teams often could do no more than view road cuts quickly in armed convoys of automobiles. Although there were a few data from Qinghai, the paucity of coverage of the desert western provinces (Qinghai, Xinjiang, and Tibet) has meant that this study omits these marginal-agriculture provinces throughout. The islands are also omitted, except for a few Hainan profiles that are included in the data on Guangdong.

The second wave, here referred to as "the 1950s," started slowly in 1950 and reached high tide in 1958–1961. For this 1950s wave, we have gleaned a few hundred typical soil profiles from several articles in Chinese, a Sino-Soviet compilation (Kovda 1959), the published but rare volume for the First National Soil Survey (1958–1961), and a special study of Inner Mongolia in 1961–1964.[5] Here again, the geographic coverage is uneven, as shown in figure 3.1b, with the most extensive coverage per unit of agricultural land area being in Inner Mongolia and Liaoning. Data from the Great Leap era (1958–1961) would naturally be suspect. Yet, as chapter 4 explains, detailed analysis of the 1958–1961 soil data does not show any optimistic bias akin to the Great Leap exhortations that distorted the national figures on production, such as the inflated grain and steel figures of that time. As far as we can tell, China's soil scientists, far removed from the political controversy of the Great Leap era, were able to maintain scientific standards.

Finally, the ambitious Second National Soil Survey of 1981–1986 is now available in a national summary volume (Institute of Soil Science 1996) and in a mixture of published and in-press volumes for regions and provinces.[6] These volumes are based on the most massive data collection in either of the two countries studied here. In fact, for many parts of China, the regional and province volumes of the Second National Soil Survey offer two kinds of soil profiles, both an "average" profile averaged over many sites and a "typical" or "representative" prototype profile taken from a single site. Either the average profile or the modal typical profile is based on a large number of individual soil profiles, so that the

Table 3.1
Numbers of soil profiles in the China soil sample, 1930s–1980s

Province	Organic matter			Total nitrogen			Total phosphorus			Total potassium			pH		
	1930s	1950s	1980s	1930s	1950s	1980s	1930s	1950s	1980s	1930s	1950s	1980s	1930s	1950s	1980s
Yunnan	3	5	105	0	5	105	3	6	105	3	3	97	5	8	109
Guangxi	0	9	120	0	11	120	0	11	120	0	6	120	51	10	120
Guangdong	0	10	131	0	11	131	0	11	130	0	9	123	10	11	128
Fujian	0	5	34	0	5	32	0	1	31	0	1	31	0	6	80
Jiangxi	3	4	108	0	4	108	2	2	108	2	0	108	9	4	106
Hunan	4	3	45	0	3	46	4	3	58	4	1	59	34	3	48
Guizhou	0	7	68	0	6	67	0	7	68	0	7	35	87	5	66
Sichuan	14	6	70	6	8	70	6	8	69	6	6	67	88	2	75
Hubei	0	2	59	0	2	59	0	1	59	0	1	59	9	3	59
Anhui	0	9	109	0	8	108	0	5	107	0	0	108	22	3	110
Zhejiang	0	2	117	0	4	115	0	0	104	0	0	71	5	2	115
Shanghai	0	0	22	0	0	22	0	0	22	0	0	23	0	0	50
Jiangsu	7	9	66	7	10	64	7	8	55	7	1	21	102	11	66
Shandong	2	2	236	0	3	236	2	2	232	2	2	129	29	15	232
Henan	0	0	94	0	1	94	1	0	88	1	0	32	15	1	70
Shanxi	8	1	64	12	1	64	12	0	64	13	0	21	18	1	61
Shaanxi	6	0	128	5	3	129	7	3	124	7	3	88	37	4	105
Gansu	2	11	53	2	19	53	0	2	52	0	2	41	34	14	56
Ningxia	0	4	47	0	4	37	0	2	36	0	3	27	0	3	50

Table 3.1
(continued)

Province	Organic matter 1930s	1950s	1980s	Total nitrogen 1930s	1950s	1980s	Total phosphorus 1930s	1950s	1980s	Total potassium 1930s	1950s	1980s	pH 1930s	1950s	1980s
Nei Mongol	5	109	170	0	104	159	0	9	166	0	8	161	10	106	165
Hebei	3	2	88	3	2	88	8	2	87	7	2	74	22	10	79
Beijing	0	2	0	0	2	0	0	3	0	0	2	0	0	3	0
Liaoning	0	0	117	0	1	117	0	1	114	0	1	107	0	0	115
Jilin	0	1	89	0	8	92	0	8	91	0	8	91	0	0	68
Heilongjiang	0	0	34	0	6	43	0	6	43	0	1	40	31	9	36
All China	57	203	2,174	35	231	2,159	52	101	2,133	52	67	1,733	618	234	2,169

Notes: Here and throughout the China chapters, "1930s" refers to soil profiles taken between 1932 and 1944, "1950s" refers to soil profiles taken between 1950 and 1964, and "1980s" refers to soil profiles in the Second National Soil survey taken between 1981 and 1986. The numbers of profiles here are sometimes less than the numbers of profiles for which the relevant soil characteristics are available in the original data sample, and are sometimes slightly greater than the number actually used in regression analysis. The slight differences in these counts arise from differences in restrictiveness for analytical purposes. For example, the counts in this table exclude histosols as well as profiles with inappropriate soil layer depths. For regression analysis some cases were omitted where other relevant variables were not available. The counts for Hebei exclude the municipalities of Beijing and Tianjin. Qinghai, Xinjiang, Tibet, and the islands are excluded from all the soil samples.

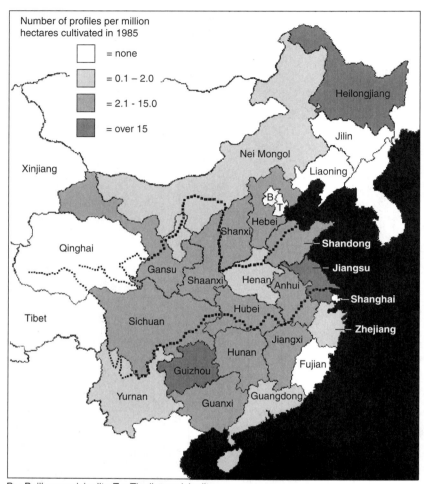

B = Beijing municipality, T = Tianjin municipality

Figure 3.1a
Soil profiles per unit of area, China pH sample, 1932–1944

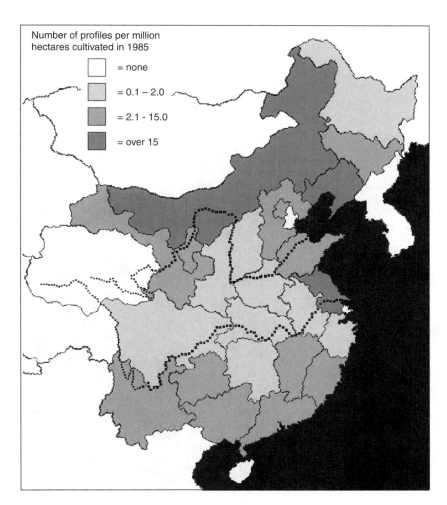

Figure 3.1b
Soil profiles per unit of area, China pH sample, 1950–1964

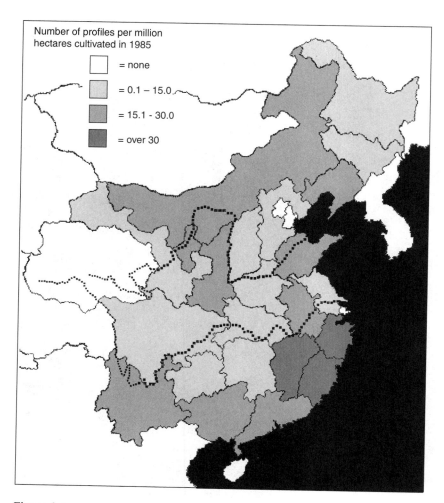

Figure 3.1c
Soil profiles per unit of area, China pH sample, 1981–1986

statistical sample for China is actually more extensive than the number of observations may imply for the 1981–1986 period, which I refer to as "the 1980s." The geography of this 1980s sample is summarized in figure 3.1c.

Thus the data are rich for China in the 1980s, but much thinner for the first two waves. We were therefore able to use only a few hundred pre-1980 observations for the whole country, with incomplete detail on land use in particular, to supplement more than 1,000 profiles for 1981–1986. The serious paucity of data for certain parts of China limits the range of conclusions that can be drawn about trends, as is noted later in this chapter and throughout part II.

Through all the periods these data cover, both the intellectual leadership and the measurement techniques in China remained much the same.[7] To determine whether the techniques used to measure pH and nutrients in the 1930s and 1950s were comparable to those used in the 1980s, we examined the original field measurement techniques. In the case of pH, the techniques seem consistent over time. The 1930s values were determined by hydrogen electrode in 1:2.5 soil-water suspension of air-dried samples that had been passed through a 2 mm sieve, and similar techniques have been followed since. For organic matter, most values in the 1930s were determined using the sulfur dioxide and potassium permanganate methods. Though these seem broadly comparable with later values, some measures in the 1930s had to be excluded from the sample as essentially noncomparable with other measures.[8]

Indonesia

Indonesia is one of the few countries with abundant soil profile data back to the early twentieth century. Even before 1910, Dutch researchers had begun to publish the basic topsoil chemistry of lands to be developed for colonization and plantations (Chin A Tam 1993). The early published samples numbered only in the dozens and measured only a few attributes. Soil profiles became increasingly frequent across the 1920s and 1930s, reaching their zenith just before World War II. The Indonesian government has continued since that time to invest in soil surveys, using procedures that appear to be consistent with those administered by the Dutch. Indeed, there is a continuity of personnel on the surveys. Much of the survey work in the 1930s and 1940s was conducted by Indonesians, and much was reported in the Indonesian language from the late 1930s on.

The resulting data are abundant and increasingly detailed. An uncatalogued archive of more than 100,000 soil profile analyses from 1923 to 1983 is currently housed in the Center for Soil and Agroclimate Research (CSAR) in Bogor. This has been supplemented by published reports on selected areas in the 1980s, plus a large and growing computerized data set since 1985. Thanks to CSAR's cooperation, this study draws on that unpublished, noncomputerized data set plus a few published data volumes for 1981–1990. Table 3.2 and figure 3.2 summarize the inventory of Indonesian data used in this study.

The early data had many omissions, however, and the research process here had to concentrate on those with the best detail. First to be cut from the sample were profiles from unknown locations (I needed at least the district, or *kabupaten*). The second cut removed those without recorded topsoil depths. After that, I eliminated profiles that gave no more than two key soil attributes (e.g., giving only P_2O_5 and K_2O but not pH or organic matter or N). Finally, the sample was again confined to mineral soils, eliminating the always small share of highly variable histosols with more than 20 percent organic matter. The resulting sample of 4,562 profiles covered pH, organic matter, and the classic nutrients more evenly than in the China sample, and for this reason table 3.2 has omitted as unnecessary the detail by soil characteristic that was included in table 3.1.

The Indonesian data set has the important advantage of allowing us to study five different types of land use separately. Chapter 7 follows separate soil trends for each of five fairly distinct types of land use system:

1. paddy (*sawah*) fields, which are generally flat, irrigated, and worked intensively, with a large share of the area devoted to rice;

2. fields concentrating on nonrice crops, which are generally upland, rain fed, and sloping (*tegalan*);

3. tree crop groves and plantations, consisting of rubber, coffee, tea, tree fruits, palm oil, teak, and cultivated bamboo;

4. fallow and grasses, covering lands lightly or previously cultivated by humans; and

5. primary forest (*hutan primer*).

Other Countries

Could the present study be replicated for other countries? As time passes, the answer will be increasingly affirmative, as several countries continue to amass large survey data archives. Many of the brightest data prospects are not, however, in the parts of the world where soil history is most urgently needed.

The top-priority countries for study would be developing countries from a variety of climates and with a variety of agricultural institutions. Although China and Indonesia would always be near the top research priority, along with India, the top global priority would be soil time series for Sahelian Africa, to follow the course of desertification and reclamation across the continent from Senegal to Somalia. Unfortunately, this area yields no historical series before the 1980s, so the necessary quantitative history is missing. The only modern-standard soil samples available from years before 1960 are from a handful of atypical sites, such as colonial white farms in Kenya. Only in the last quarter of the twentieth century have soil profiles begun to accumulate in Sahelian Africa, and most of these are from experimental plots. The soil history of Sahelian Africa must therefore await a combination of suggestive satellite data, core samples, and profiles from the twenty-first century.

India has the makings of a rich quantitative soil history dating from the 1950s. Although India's usable soil profiles go back as far as 1934, only in the 1950s did the broad accumulation and publication of profile data begin (Indian Council of Agricultural Research 1957; Raychaudhuri et al. 1963). These early published profiles could be supplemented with additional early profile data archived by the Indian Council of Agricultural Research in Nagpur. In the 1980s India, like China, undertook a massive national survey, the reconnaissance survey supervised by the National Bureau of Soil Survey and Land-Use Planning. In addition, more detailed field surveys were conducted in the 1980s by four laboratories in Delhi, Nagpur, Calcutta, and Bangalore. When extended into the twenty-first century, the Indian soil database should be quite rich.

Two other countries have very long-standing soil science information. One has abundant data, and the other has many fewer, despite its soil science tradition. The United States has abundant soil profile data covering the length of the twentieth century, and for the period since the 1930s

Table 3.2
Numbers of soil profiles used in this study by province and decade, Indonesia, 1923–1990

Province	1923–30	1931–40	1941–50	1951–60	1961–70	1971–80	1981–90
Aceh		18			5		
North Sumatra		40	24	48	46		
West Sumatra		19		49			
Riau		44		39	15	40	45
Jambi		36	30	8			
South Sumatra	4	33	7	27	47	60	35
Bengkulu		11		9			
Lampung		25		43	52	44	33
West Java (including Jakarta)	12	176	136	174	124	129	123
Central Java (including Jogjakarta)	32	127	85	143	215	95	56
East Java (including Bali, Madura)	19	168	156	130	98		
West Kalimantan		46		41	4	36	134
Central Kalimantan		12	14	24	2	45	
South Kalimantan		32	14	16	3	16	
East Kalimantan		12		50		21	71
North Sulawesi	4	57		5			
Central Sulawesi			7		12		26
South Sulawesi	7	56	49	3	54		
Southeast Sulawesi		7	71		4	59	

Table 3.2
(continued)

Province		1923–30	1931–40	1941–50	1951–60	1961–70	1971–80	1981–90
Nusa Tenggara (East and West)			4	51	22	38		51
Maluku			5	25	36	1		
Irian Jaya			31	36				
Totals								
Indonesia	4,562	78	966	698	879	708	571	662
Sumatra	936	4	226	61	223	165	144	113
Java, Bali, Madura	2,230	63	471	377	447	437	224	
Kalimantan	593	102	28	131	9	118	205	
Sulawesi	503	11	127	120	20	58	85	82
East (NT, MK, IJ)	300	40	112	58	39	51		

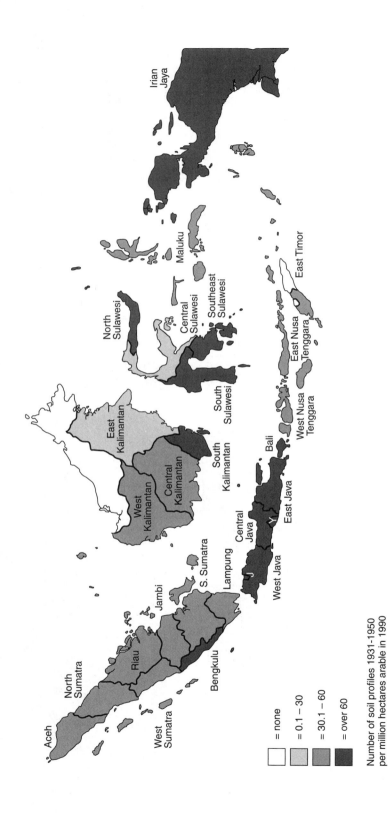

Number of soil profiles 1931-1950
per million hectares arable in 1990

☐ = none

▨ = 0.1 – 30

▨ = 30.1 – 60

■ = over 60

Figure 3.2a
Soil profiles per unit of area, Indonesia, 1931–1990

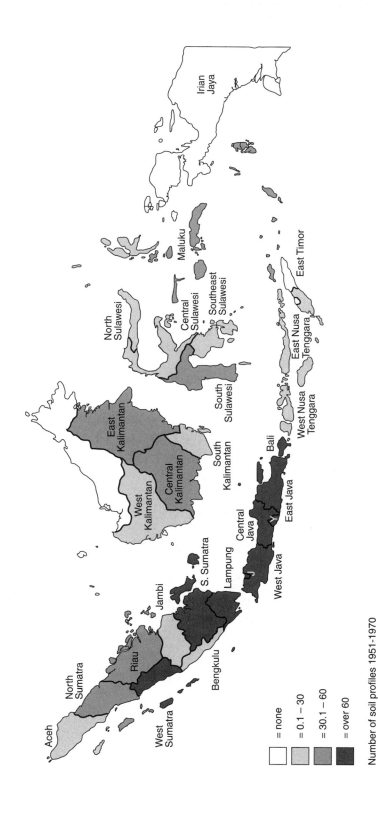

Number of soil profiles 1951-1970
per million hectares arable in 1990

= none

= 0.1 – 30

= 30.1 – 60

= over 60

Figure 3.2b
Soil profiles per unit of area, Indonesia, 1951–1970

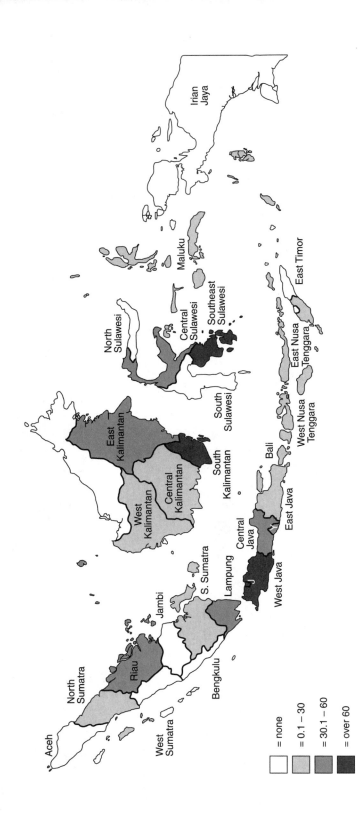

= none

= 0.1 – 30

= 30.1 – 60

= over 60

Number of soil profiles 1971-1990
per million hectares arable in 1990

Figure 3.2c
Soil profiles per unit of area, Indonesia, 1971–1990

probably the richest abundance of survey data of any country in the world. Russia/the Soviet Union, a pioneer in soil science, has studies of soil chemistry and groundwater extending back to the Czarist era. Yet for the United States, the issue of soil trends lacks the urgency it has in developing countries, and the availability of abundant long-run soil data for Russia is in doubt. More promising, therefore, as a basis for conclusions about soil trends over much of the world would be to concentrate on the rich data sets and pressing conservation issues of India since the 1950s and—above all—China and Indonesia.

Which Soil and Site Characteristics Are Covered, and Which Are Not?

Soil Characteristics   Before the 1980s the soil profile data from China and Indonesia covered a varying list of soil characteristics, and a long history is possible only for certain characteristics that were consistently measured. The most consistent coverage is of pH, organic matter (OM), and three classic macronutrients—total nitrogen (N), total phosphorus (P, or P_2O_5 content), total potassium (K, or K_2O content), and OM—along with calcium (CaO; for Indonesia), soil texture shares (percentage sand, percentage clay), parent material, and a soil classification. Physical composition (including aluminum-iron-silicon breakdowns) is generally available for the 1970s and 1980s but is not used here because it is only intermittently available for earlier decades. The data are presented for different perceived soil "horizons" down to various depths. Among the desirable information not consistently included in the raw data are the "available" levels (as opposed to total levels) of the macronutrients noted above and of three other macronutrients (calcium, magnesium, sulfur), or any coverage of micronutrients, or of cation exchange capacity, or of salt content. What we are offered for several decades is a list of variables that is satisfactory, though not ideal.

Although the available classic indicators fall short of the list of minimum data needed to quantify soil quality (Pierce et al. 1983; Rijsbergen and Wolman 1984; Doran et al. 1994), they figure prominently in such a list. Since we are interested in one particular dimension of soil quality— namely, agricultural productivity—it is fortunate that those who chose which data to gather and present over the decades shared the same priority. Thus the few indicators we stress here predict most of the soil's effects on yields fairly well. In particular, bulk density, a good predictor of plant growth, can be well approximated just by knowing a site's texture and organic matter (Rijsbergen and Wolman 1984).

The coverage of soil characteristics is more uneven for China than for Indonesia, as already hinted by the contrasting presentations in tables 3.1 and 3.2. The data for Indonesia cover pH, OM, N, P, and K with similar fullness, so that the numbers in table 3.2, which represent the number of profiles used for at least one characteristic, are not much different from the sample sizes used for each individual characteristic.

The data coverage for China calls for more discussion, however. The unevenness of the data coverage of soil characteristics requires us to choose carefully any conclusions made from that data about trends in China. It is necessary to identify which combinations of place, time, and soil characteristics are not well enough documented to establish trends. Table 3.1 and figure 3.1 have identified some thinly covered provinces in the various time periods they illustrate. Rather than provinces, however, the relevant "places" to consider here are those are used as statistical test areas in chapters 4–6. Some of the soil samples used in those chapters consist of a particular agricultural region and some consist of the totality of north China or of south China. As a preview of trend conclusions that *cannot* be drawn, here is a quick summary of the coverage of the main soil characteristics in three eras for the whole of north China and the whole of south China:[9]

1. The 1930s, meaning 1932–1944, yields the following numbers of soil profiles for each main chemical characteristic:

	pH	OM	N	P	K
North	209	25	22	24	25
South	412	32	9	22	22

The sampling of pH is far more robust than that of organic matter and the NPK macronutrients. Nitrogen has the poorest coverage for the 1930s. Certainly no trend conclusions involving the 1930s should be made about nitrogen in the south. To posit a trend in any other nutrient one must show that the trend appears quite pronounced or is supported by evidence from outside the sample.

2. The 1950s, meaning about 55 profiles for all China 1950–1958, plus 129 from the First National Soil Survey taken 1958–1961 and 116 profiles from a special study of Nei Mongol 1961–1964, yielded:

	pH	OM	N	P	K
North	178	146	142	32	29
South	69	56	60	49	27

Here the coverage looks better balanced in the north, though table 3.1 shows that much of the northern sample's coverage of organic matter and nitrogen is from Inner Mongolia. Almost every province south of the Chang Jiang (Yangzi River) is reasonably represented, as is souther Jiangsu, but there are virtually no 1950s data for Sichuan, Hubei, Anhui, or Zhejiang. In both north and south, the coverage of P and K is thin enough to prevent involving the 1950s in any discussion of trends without supporting evidence from elsewhere.

3. The 1980s, meaning the Second National Soil Survey taken 1981–1986, yielded these numbers of profiles for each main chemical characteristic:

	pH	OM	N	P	K
North	1119	993	976	961	697
South	1135	1135	1125	1111	993

Clearly, the 1980s data will not present binding sample size constraints on conclusions about trends. Every case involving constraints on drawing trend conclusions will arise from the paucity of data for either the 1930s or the 1950s.

We focus on the topmost (A) horizons of mineral soils. We have excluded organic soils (bog or peat soils, or those with OM > 20 percent) and deep geological profiles (where the A horizon extended deeper than one meter). The concept of a "topsoil" layer also warrants a quite specific definition. To focus on comparable root zone "A horizons" (a specific topsoil layer), instead of measuring surface debris, the sample includes only horizons including the depth of six centimeters from the surface. That is, the layers studied here are only those that begin within five centimeters of the surface and end six or more centimeters from the surface—essentially, the upper root zone.

Site Characteristics   Making the data from soil profile sites serve as suitable representatives for the vast expanses of untested locations requires finding the statistical patterns that best link soil characteristics with characteristics of the sampled sites. Several variables are key here, and most of them are available. The first is simply the soil type as classified by the nation's soil scientists, whose classification system is often based on criteria that maximize the similarity of characteristics among the sites in each soil class. Whether the nation's soil scientists use their own classification system or opt for an international or U.S. Department of Agriculture (USDA) taxonomy is of only secondary importance. One way or the other, the soil classification system is designed to do what we want it to do, namely to maximize the similarity of soil characteristics, including soil chemistry, of the soils in each group. Fortunately, useful soil classifications are available for both China and Indonesia. Even when the soil profile source data do not provide them, they can often be inferred from the site location.

Two other key determinants of soil chemistry can also be inferred from location when not given by the soil data sheets themselves. One is the parent material (volcanic soils, alluvial, forest and bog soils, etc.). The other is climate. Data on these variables were available in nearly all cases.

Land use is another key attribute for determining the soil chemistry of each site and for inferring how humans are affecting the soil over time. Data on this site variable are much better supplied for Indonesia than for China. As already noted, Indonesia's data allow us to pursue separate soil trends on five kinds of land use systems since the 1930s, including a virgin forest baseline. China's soil data, however, often lack a specific identification of land use. Although this remains a serious shortcoming, it has been minimized with the help of some indirect indicators of land use that fit a particular site's environment if not the exact site where the sample was taken. First, chapter 4 divides China into agricultural regions, so that we know the general crop mixture for lands known to be tilled. Second, we usually know whether the land is currently being tilled, especially for the 1980s data.

The one key site attribute that is missing throughout the data set is the earlier land use history of the tested site. Seldom do we know how a given site was cultivated five or ten years earlier, unless it is virgin forest or has some special designation revealing its past, such as "reclaimed swamp" in China or "formerly paddy" (*bekas sawah*) or "formerly shifting cultivation" (*bekas ladang*) in Indonesia. This data gap poses a serious challenge when we try to interpret the effects of particular kinds of land use.

Since the sampled sites are in general more similar in their unknown past use than in their known present use, the differences in their soil chemistry probably underestimate the effects of permanently practicing each kind of land use. To take a Chinese hypothetical example, similar to an Indonesian example I use in chapter 7, consider the differences we are likely to find between the nitrogen contents of cotton fields and fallow. In a countryside where some cotton land was once fallow, and most fallow was once planted in cotton and other crops, comparing their nitrogen content may well suggest that land use makes no difference to the soil nitrogen. Suppose that continuous cotton cultivation lowers nitrogen from an index of 100 to an index of 60 steadily over ten years, and leaving the same former cotton fields fallow for eight years returns the nitrogen steadily from 60 to 100. Then if we sample all years in this cycle evenly, the average nitrogen will be 80 on both the cotton lands and the fallow lands. Yet it would be a mistake to conclude that land use makes no difference. Our inability to see the land use dynamics of the individual site is a handicap, and no conclusions should be reached which are very sensitive to this point.

A Statistical Strategy for Separating Time from Space

The Challenge: Comparing "Comparable Places"
The most natural way to hold place constant is to compare repeated soil samplings from the same site. But aside from the few long-run experimental plots, which cannot reveal what real farm populations do to the soil, same-plot sampling has not occurred. Soil profiles are almost never taken from the same place, even from the same 100 hectares, more than ten years apart, except on long-term experimental sites like the Rothamsted or Morrow plots. Rather, historical comparison requires knowing that two different sites are sufficiently the same in their broad classes of soil attributes and have the same relationship to the mean values of the variables of interest (pH, OM, and the NPK macronutrients) within each

set of "same" soils in different time periods. These requirements for "sameness" over time roughly match the taxonomic goals of any regional soil survey taken in a single time period. Soil scientists believe that soils can be meaningfully clustered according to certain attributes that are relevant for agriculture as well as for geological and other scientific purposes, and we make use of such clustering here.

But are those soil profiles ranging from the 1930s to the 1980s really comparable? Or do differences in their original sampling purposes and in the soils selected preclude our knowing what has happened to soils of given attributes over this half century? Different samples can indeed reveal trends in similar soils, but only after a sound statistical procedure has been followed.

Widespread soil degradation or soil improvement should show up as a downward or upward trend in soil parameters when we hold other things equal in our comparison of profiles from different historic eras. To hold other things equal and to reveal systematic changes in soil chemistry relative to a particular kind of soil at a particular time and place. Such comparisons run through the chapters ahead. The different soils of north China are compared with a particular kind of tilled alluvial soil in the Huang-Huai-Hai plain in the 1980s. Soils in south China are compared with tilled alluvial soils of either the Sichuan basin or the north Chang Jiang plain. For Java soils, the comparison standard is alluvial paddy (*sawah*) soils near Semerang in Central Java in 1990. And the diversity of Indonesia's outer island soils is viewed against the background of alluvial-soil paddy near Kisaren, North Sumatra, in 1990.

The Generic Equation

To hold the right other things equal requires an equation and some regression analysis. This section describes the basic equation for sorting out the attributes of place and time that determine soil pH, organic matter, nitrogen, phosphorus, and potassium. Specifically, we will explore the patterns revealed by this generic regression equation

$$Y = a + bH = cX + dS + e,$$

where

Y is pH or $\ln(OM)$, or $\ln(N)$, or $\ln(P)$, or $\ln(K)$;

a, b, c, and d are sets of coefficients;

H, or "history" variables are a vector of historical shift variables, often differentiated by region;

X, or place variables are a vector of site attributes that vary little over time (parent materials, soil classes, texture, terrain, precipitation, agricultural regions, land use, and other landscape features, and depth of the A-horizon);

e = random error; and

S represents biases in the data sources.

My use of logarithms for organic matter and NPK content levels in formulating the equation stems from the finding that using absolute percentage shares caused skewness in the error term, violating normality. Using logarithms to correct for this error skewness has the desirable side effect of giving greater explanatory weight to differences in independent variables at critically low levels of nutrients—for example, greater weight to an independent variable's ability to distinguish a 1 percent level of organic matter from a 2 percent than to the differentiation between a 14 percent and a 15 percent level.

The ability to separate human from nonhuman effects is crucial to this equation. Fortunately, distinguishing human from nonhuman effects is largely a matter of time versus space here. A sample spanning a few decades can distinguish human impacts from other effects. Several decades is a long enough time period for human management to show a clear effect, yet one short enough for soil changes to owe little to nonhuman changes, even to climate changes and volcanic activity. Hence the time trends summarized by *H* in the equation are attributable to human influences (as long as the equation is correctly specified). The spatial *X* variables can be thought of as representing nonhuman influences, since even spatial differences in human treatment of the land are largely responses to nonhuman physical variables.

The historical *H* variable needs to be something more flexible than just a linear time trend, since we have no way of knowing in advance whether a particular trend has been steady over the decades. For China, where the data are grouped into three eras, we simply compare each of the two earlier eras separately with the 1980s, so there is an *H* variable to represent any observation that came from the 1930s (1 if from the 1930s, 0 otherwise), and another for the 1950s. For Indonesia, the continuous flow of soil profiles over half a century calls for chapter 7's more complex set of *H* terms to represent human influences over time.

In Search of Errors

Several kinds of potential "errors" call for careful attention in the type of exploration of how time, place, and pure randomness determine differences and changes in the soil that the last section's equation attempts to make possible. Aside from being alert to the need to edit the raw data for simple mistakes, some considerable care needs to be directed at these specific kinds of errors:

1. The random noise term *e* in the generic equation, potentially complicated by

1a. heteroskedascity, and

1b. censoring of the dependent variable.

2. Systematic bias in the data themselves, because of a bias in our original soil-measuring source, the *S* variables. The main biases to consider are

2a. shifts in the whole purpose and mood of soil sampling in China from one era to another,

2b. data biases for parts of China in the 1980s,

2c. possible shifts in the laboratory measurement technique in Indonesia, and

2d. data biases for parts of Indonesia.

Let us discuss each type of error briefly before we encounter them again and treat them more fully in the chapters ahead.

   1a. It is uncertain whether the error term *e* would be a well-behaved homoskedastic disturbance with the same variance for all observations. The sources may differ in the amount of sampling and averaging that they put into each "representative" soil profile, suggesting that data from different sources have different error variances. Here again, we enlist the help of the source bias (*S*) variables. Using the China samples as the laboratory, we posit that the absolute value of *e* depends on a well-behaved error term (*u*) and on the source variables in a multiplicative way:

$|e| = |u| \cdot r(S)$.

To estimate r, we use separate regressions explaining the absolute errors of a preliminary ordinary least squares regression in terms of the Ss. Dividing all other variables by the predicted value of r yields a modified equation that should have an error term (u) with a desirably uniform variance:

$Y/r = a/r + b(H/r) + c(X/r) + d(S/r) + u$

Performing this transformation for the China data turned out to make little difference in the main conclusions. And since inspections of the Indonesian data found little basis for expecting much heteroskedasticity, no such data transformation seemed necessary for Indonesia.

1b. A different procedure is in order when the dependent variable is bounded on one side. The convenient variable pH is a bit off the mark as a variable of interest in spotting trends toward, or away from, soil degradation that is relevant to plant growth. Rather we are interested in only the more extreme ranges of pH. For north China, alkalinity—the extent to which pH exceeds, say, 8.0—is more important than just pH. A soil pH of 9.0 rather than 8.0 constrains crop choice more severely than a difference in the range from 6.5 to 7.5. Similarly, for south China or Indonesia, acidity, which is defined as the shortfall of pH below 6.0, is more worth monitoring than differences in pH across the neutral range between 6.0 and 8.0. Thus we are interested not in pH itself, but in those two variables that are limited at zero:

Alkalinity = max (pH − 8.0, 0), and

Acidity = max (6 − pH, 0).

The fact that these variables press against the zero limit means it is unwise to use ordinary least squares as a way of viewing the error term e. Here we will use a tobit procedure (see Maddala 1983) to estimate either alkalinity or acidity as a combination of the probability of its being positive and its magnitude if positive.

2a. The source bias (S) variables are included because some sources may have had systematic errors with nonzero means. Wherever we suspect such a bias in a particular source, we add an S variable that equals one for every profile from that source. To see why this is necessary, suppose that China's 1930s soil testers sought to sample and analyze the most troubled and unproductive soils, whereas the 1950s samplers sought to show off the best soils. Without a correction for this source bias, comparisons of data between those two eras could mislead us into thinking there was a general improvement in China's soils when there was none. The example represents a real fear with which the present author approached the part of China's 1950s data set coming from the first National Soil Survey (1958–1961). It might have focused on sampling the most fertile lands, lands that would best support the intense cropping called for by Chairman Mao in those years of the Great Leap Forward. We report on some of our source bias investigations below.

2b. Within the 1980s, my research team's reading of the Chinese provincial soil summaries suggested that a few provinces may have reported data from better lands whereas a few others gave more attention to poor lands. We harbored suspicions about the puzzling high levels of some nutrients in Yunnan and Guizhou in the 1980s, for example. One way to check these suspicions became available in 1996, when the Office of the National Soil Survey released its national summary volume for the second National Soil Survey. Given that our data set (gathered in 1993–1994) contained many rough preliminary returns, it should have been possible to test whether the lack of access to the edited final product caused any serious distortions in the data we used from the second National Soil Survey. As it turned out, the data in the summary volume confirmed the high levels of nutrients in Yunnan and Guizhou.

2c. In the case of Indonesia, it is similarly important to explore possible biases by place and time period. Chapter 7 reports some evidence of both. A few of the districts reported peculiar-looking results, and the data for the whole era 1974–1986 seemed to have oddly lower values of phosphorus and potassium for the whole of Indonesia. The work underlying chapter 7 quantified these biases. The 1974–1986 anomaly, at least, appears to be the result of shifts in laboratory measurement technique.

Despite all the concerns about the data that must be addressed, the soil profiles of China and Indonesia can yield meaningful conclusions, even some unanticipated conclusions, about soil use patterns and soil quality in the two regions. The precautions described in this chapter limit, but also strengthen, the conclusions of parts II and III.

II

China

4
Soil Changes in China since the 1930s

Our soil profile data for China from the 1930s through the 1980s can shed light on previously unknown trends in soils for given site characteristics. This chapter uses the framework and the precautions of chapter 3 to establish some clear overall patterns for China's main agro-climatic regions. We will find a trend toward degradation of some dimensions of soil quality in some regions. In other dimensions we will find signs of improvement, specifically:

• Organic matter and nitrogen seem to have been depleted in the more intensely worked lands to the east, at least between the 1950s and 1980s and perhaps earlier.

• Total phosphorus and total potassium have not been depleted. On the contrary, they accumulated remarkably throughout all the parts of China for which we have a fair number of soil profiles from before the 1980s.

• The trends in soil pH were mixed. There is no sign of alkalinization in the north as a whole. Nor has soil acidification worsened monotonically in the south, as some had feared in the 1930s. Acidity abated between the 1930s and the late 1950s, though part of this improvement was reversed in some areas after the 1950s.

• The topsoil layer appears to have become thinner in the half century from the 1930s to the 1980s, as fears of erosion would predict. Yet a closer look reveals that what looks like a thinning of the topsoil was probably due to a change in the definition and purpose of defining soil layers, not to any clear change in the thickness of the A horizon as defined by today's soil scientists.

Before discussing these trend results and interpreting them, we must turn first to some physical background variables that will play key roles in distinguishing the influences of time and space in China's soil chemistry.

Physical Background

As chapter 3 has already underlined, one cannot judge the effects of human activity over time without having given nonhuman spatial variables their due. The physical geography of soils varies with parent material, soil class, terrain, climate, and land use. The last two of these physical dimensions call for further explanation here.

The two dimensions of climate that are obviously crucial to agriculture—namely, precipitation and temperature—happen to be strongly correlated in China. An axis running northwest from subtropical eastern Guangdong to semidesert Gansu defines the main variation in both average rainfall and average annual temperature, as a comparison of figures 4.1 and 4.2 suggests.[1] An axis running across the map from southeast to northwest marks out a gradient from wet and warm to dry and cold. This pattern is less convenient for our statistical inquiry than it is for the student of geography. It yields a strong correlation of .80 between precipitation and temperature in a sample of 2,032 counties across China, making it difficult to sort out the separate effects of precipitation and temperature as influences on China's soil chemistry.[2] Although it might seem wrong to omit temperature and altitude in the analysis that follows, their influences on soil chemistry are already well represented by precipitation, precipitation squared, soil classes, and landscape, and so this is the course that has been taken.

Land use is the other physical background variable that calls for comment. It is important both as a way in which humans affect the soil and as a way of distinguishing between lands affected by humans and those that remain unaffected. In China, the immediate land use is unknown for a large share of all profile sites, both in the 1980s and earlier. To alleviate this lack of information somewhat, it helps to know how agricultural sites in the area tend to be used.

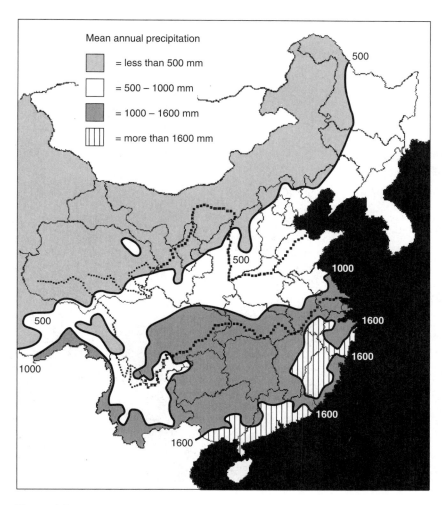

Figure 4.1
Average precipitation in China
Source: Institute of Soil Science 1986.

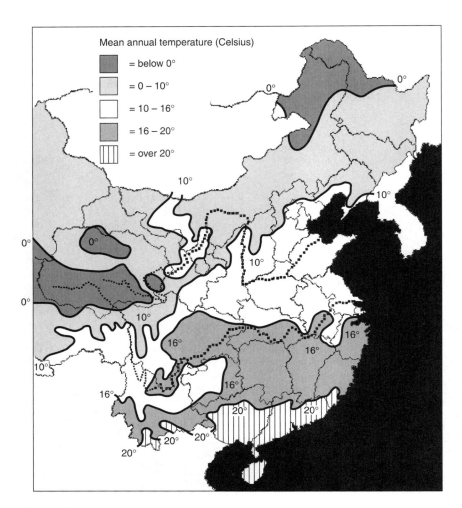

Figure 4.2
Average temperatures in China
Source: Institute of Soil Science 1986.

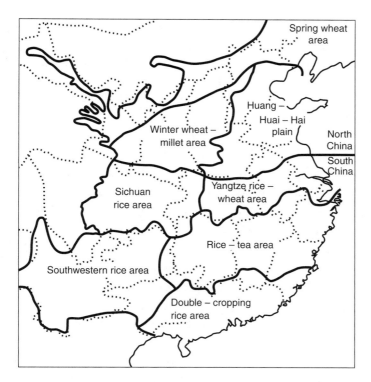

Figure 4.3
Main agricultural regions of China, as envisioned at the University of Nanking in the 1930s
Source: Based on Buck 1937, p. 27.

   For the purposes of its discussion, this book divides north China from south China at the Qin Ling mountains and the bottom of the Huai river basin. To regionalize China's diverse agricultural landscape further, figure 4.3 uses the crop regions envisioned at our 1930s starting point by the Nanking University research team. Thus the north consists of the spring-wheat region (including the northeast provinces), the winter wheat–millet region, and the winter wheat–sorghum region, which today is called the Huang-Huai-Hai plain. The south consists of the Sichuan basin, the Chang Jiang (Yangzi) plain, and everything farther south. The western hinterland of desert, steppe, and the Himalayas is excluded, as are the islands. Which of these regions a particular soil profile site is

located in, when combined with the knowledge as to whether the site was
pasture or land that is tilled at least once a year, offers a useful set of clues
about the kind of land use at a particular site, in the absence of actual
recorded data.

The Resulting Equations

Tables 4.1–4.4 summarize the determinants of soil nutrient levels, organic
matter, and pH in the topsoils of China since the 1930s. As much as is
practicable, the regressions are regionalized, to avoid the implausible
assumption that any single equation fits all of China. The upper half of
each table depicts the roles of time, region, and population. Before
turning to these key influences, we should first note that the lower half of
each table merely counts some extra variables instead of displaying all
their coefficients, in order to save space. The full equations in Appendix
A show that the extra coefficients yield plausible patterns. For example,
levels of both organic matter and of nitrogen are higher in mountainous
areas than on tilled fields and higher on tilled fields than on wastelands.
Levels of both are also higher in lacustrine deposits, and lower in quaternary red clay and in loess, than in the average alluvial soils. They are also
higher the shallower the midpoint of the A horizon. The overall explanatory power of every equation is highly significant.

The random-error terms received two kinds of special treatment in
these equations. First, as previewed in chapter 3, it is important to
explore the possibility that the error variance might differ across subsets
of the data, since they come from different original data sources. This
possibility is allowed for by a generalized least squares procedure, de-
noted by the "GLS" in some of the equations in tables 4.1–4.4.[3] Using
GLS instead of ordinary least squares made little difference, however.
Accordingly, the extra effort was avoided in a couple of small-sample
cases here, and it will be avoided in part III's regressions on Indonesian
data, for which I will revert to using ordinary least squares. Second, as
also discussed in chapter 3, a tobit procedure was used for acidity
(defined as any shortfall of pH below 6) and for alkalinity (defined as any
excess of pH above 8), because acidity and alkalinity, which are more
relevant to plant growth than the continuous pH variable itself, are

bounded by zero, violating conventional parametric assumptions about the random-error term. Although acidity and alkalinity are the relevant measures of soil reaction for agronomic and economic purposes and will be used in chapters 5 and 8, they are represented here only by two acidity equations in table 4.4.[4] Equations explaining simple pH are shown in tables 4.3 and 4.4 to accommodate readers' easier familiarity with pH and with standard least-squares equations. The patterns of determinants of pH are similar to those of acidity or alkalinity, in any case.

Armed with the historical-shift coefficients from Tables 4.1–4.4, we can now explore and interpret the soil trends for tilled soils in the different regions of China.[5] We turn first to the main agricultural regions of the north.

Soil Trends in North China

The historical trends on which we focus here are summarized for north China by the tilled-field averages in table 4.5 and figure 4.4.[6] Each average holds constant a host of physical and data source variables (all the Xs and Ss in Chapter 3's generic equation) and is to be interpreted as showing the average for the fixed mixture of soil conditions prevailing for each type of site in any county within that region in the 1980s.

Northern Trends in Macronutrients

Over most of north China, there was no significant trend in changes in the level of organic matter or of total nitrogen between the 1930s and the 1980s. Figure 4.4 shows no clear trend for 1930s–1950s and a likely decline thereafter, though the fall after the 1950s fails to show statistical significance for north China as a whole.

For two major regions, however, we have evidence of a substantial decline in organic matter and nitrogen from the 1950s to the 1980s. The first is Nei Mongol, whose soil profiles for 1961–1964 make up most of the "1950s, spring wheat" area listed in table 4.5 and figure 4.4. The second is the great Huang-Huai-Hai plain. In both regions, the statistical significance is only marginal, but the implied decline is large. By contrast, there were no significant losses of organic matter and nitrogen for the Shaanxi-Shanxi area (the "winter wheat–millet" region of figure 4.3) or

Table 4.1
Determinants of soil nutrient levels in north China, 1930s–1980s

Independent variables	Equation numbers and dependent variables							
	(1) ln (percentage organic matter)		(2) ln (percentage nitrogen)		(3) ln (percentage phosphorus)		(4) ln (percentage potassium)	
	Coeff.	|t|	Coeff.	|t|	Coeff.	|t|	Coeff.	|t|
Time period and region (relative to 1980s, same region; – = positive growth up to 1980s)								
1930s, winter wheat–millet	-0.059	(0.37)	-0.128	(0.80)	0.127	(1.10)	-0.057	(0.68)
1930s, spring wheat	0.472	(1.89)[a]	-0.253	(0.78)				
1930s, Huang-Huai-Hai	0.185	(0.68)	0.552	(1.71)[b]	-0.202	(1.63)[b]	-0.427	(3.53)**
1930s, desert-steppe	-0.655	(0.90)	-0.741	(1.05)				
1950s, spring wheat	0.142	(1.82)[a]	0.065	(0.88)	-0.398	(2.54)*	-0.274	(2.57)*
1950s, winter wheat–millet	0.140	(0.59)	0.063	(0.36)				
1950s, Huang-Huai-Hai	0.215	(1.15)	0.369	(2.05)*	0.263	(1.64)[b]	0.051	(0.49)
1950s, desert-steppe	-0.405	(0.86)	-0.512	(1.17)				
Region (relative to the Huang-Huai-Hai plain, same time period)								
Spring-wheat region	0.337	(3.82)**	0.351	(4.01)**	0.247	(2.71)**	0.066	(1.41)
Winter wheat–millet	0.442	(5.71)**	0.487	(6.72)**	0.109	(1.76)[b]	0.101	(2.31)*
Desert-steppe region	0.912	(1.94)[a]	0.971	(2.21)*	0.124	(0.44)	0.067	(0.38)
Northeast provinces	0.116	(1.34)	0.059	(0.76)	-0.197	(1.79)[b]	0.019	(0.46)
Density of human employment in that county in 1982								
ln (agricultural), T	-0.105	(4.65)**	-0.087	(4.07)**	-0.024	(1.17)	0.007	(0.49)
ln (nonagricultural), T	0.078	(3.57)**	0.045	(2.07)*	0.011	(0.53)	0.008	(0.62)

Table 4.1
(continued)

Independent variables	Equation numbers and dependent variables							
	(1) ln (percentage organic matter)		(2) ln (percentage nitrogen)		(3) ln (percentage phosphorus)		(4) ln (percentage potassium)	
	Coeff.	|t|	Coeff.	|t|	Coeff.	|t|	Coeff.	|t|
Numbers of other independent variables in the same equations								
More density variables	2		2		2		2	
Landscape variables	8		8		8		8	
Parent materials	9		9		9		9	
Soil classes	24		24		24		24	
Precipitation variables	2		2		2		2	
Others (texture, depth, source bias)	9		8		10		9	
Adjusted R-squared	0.602		0.787		0.733		0.466	
Dependent variable mean	0.372		−2.584		−3.218		0.765	
Standard error of estimate	0.559		0.531		0.548		0.317	
Least-squares type	GLS		GLS		GLS		GLS	
Number of observations, degrees of freedom	1,164 1,095		1,140, 1,072		1,017, 952		751, 687	
Observations from 1930s	25		22		24		25	
from 1950s	146		142		32		29	
from 1980s	993		976		961		697	

Table 4.1
(continued)

Notes: Fuller versions of these regressions are given in appendix A.

The nutrient levels refer to total levels, not available levels.

The agricultural regions are those shown in figure 4.3. The northeast provinces of Liaoning, Jilin, and Heilongjiang are counted as part of the spring-wheat region. This study omits the desert far west (Qinghai, Tibet, Xinjiang) and the islands.

In (agricul. density), T = the natural logarithm of the number of persons employed in agriculture per square kilometer in that county (*xian*) in 1982. The T indicates that this was a case in which the soil profile site was explicitly described as tilled. The density refers to 1982, whatever the date of the soil profile, since county level population densities were not available for earlier years.

In (nonagricultural), T = the natural logarithm of the number of persons employed in nonagricultural pursuits, per square kilometer in that county in 1982. Again, T means the profile site itself was tilled.

The additional two human-density variables are the same as those just defined, except that no landscape description was given for the soil profile site itself.

The binary variables for profile site landscape descriptions other than tilled were grass, marsh and lowlands, waste or uncultivated, hillside, steep slope, mountains, and NL (no landscape given).

The nine parent material categories (compared to alluvial deposits) are weathered crystalline rock, weathered clastic sedimentary rock, weathered calcareous rock, diluvial deposits, quarternary red clay, lacustrine deposits, coastal marine deposits, loesses, and aeolian sands. These were taken from the 1986 *Soil Atlas* if they were not given in the soil profile data source.

The twenty-four soil classes are large-group classes taken from the agricultural atlas of China. All other classes are compared with *chao tu fluvo-aquic*, a common alluvial soil class.

The two precipitation variables are average precipitation (meters/year) and its square.

The texture of the profile's A horizon is summarized by the quadratic combination of five percentage shares: sand (and gravel), sand squared, clay, clay squared, and sand times clay. The silt percentage and its square are thus implicit.

Depth is represented by the variable *shallow*, which equals 1/(depth of the profile midpoint).

The *source bias* binary variables represent data sources that were suspected of being biased relative to most other sources used in the sample. They included a few provinces in the 1980s (since the 1980s data were gathered by province), along with Kovda's book covering the 1950s, and Thorp's pH map for the 1930s (from which a few pH observations were borrowed).

Table 4.1
(continued)

The dependent-variable mean and standard error are those of the implicit original equations, not the values divided by rho in the generalized least squares (GLS) equation.

I used GLS whenever an earlier regression showed that errors from an ordinary least squares (OLS) regression were significantly related to differences in data sources.

The numbers of profiles cited in the tables and the text differ according to their specific purpose. Those mentioned when the data are introduced in the text are the numbers of observations usable in any of the three tables. Those summarized in the tables are slightly smaller, because they exclude a few profiles from the desert-steppe region. The sample sizes here include the desert-steppe region, but exclude profiles lacking certain independent-variable data.

Table 4.2
Determinants of soil nutrient levels in south China, 1930s–1980s

Independent variables	Equation numbers and dependent variables							
	(1) ln (percentage organic matter)		(2) ln (percentage nitrogen)		(3) ln (percentage phosphorus)		(4) ln (percentage potassium)	
	Coeff.	|t|	Coeff.	|t|	Coeff.	|t|	Coeff.	|t|
Time period (relative to 1980s; − = positive growth up to 1980s)								
1930s, all south China	−0.2853	(1.60)	−0.1286	(0.64)	−0.2048	(1.40)	−0.9951	(5.61)**
1950s, all south China					0.0618	(0.48)	−0.3321	(2.44)*
1950s, southwest China	−0.4852	(3.61)**	−0.1528	(1.09)				
1950s, other south China	0.0406	(0.35)	0.1961	(1.79)^b				
Agricultural region (other than the north Chang Jiang plain)								
Desert-steppe region	0.2553	(0.74)	0.1930	(0.61)	0.5164	(1.76)^b	0.0652	(0.23)
Sichuan basin	−0.0435	(0.57)	−0.0830	(1.16)	−0.0310	(0.36)	−0.2153	(3.37)**
South Chang Jiang plain	0.0554	(0.94)	−0.0024	(0.04)	−0.0650	(0.85)	−0.1529	(3.03)**
Double-cropped rice	0.1165	(1.54)	−0.0457	(0.58)	0.1346	(1.56)	−0.3098	(3.24)**
Southwestern rice	0.4365	(5.14)**	0.3915	(5.26)**	0.2516	(2.95)**	0.5268	(4.83)**
Densities of human employment in each type of county in 1982								
ln (agricultural), T	−0.0728	(2.71)**	−0.0560	(2.58)*	0.0054	(0.22)	−0.0240	(1.07)
ln (nonagricultural), T	0.0625	(2.64)**	0.0581	(2.75)**	0.0565	(2.17)*	−0.0102	(0.49)

Table 4.2
(continued)

| | Equation numbers and dependent variables | | | | | | | |
| | (1) ln (percentage organic matter) | | (2) ln (percentage nitrogen) | | (3) ln (percentage phosphorus) | | (4) ln (percentage potassium) | |
Independent variables	Coeff.	|t|	Coeff.	|t|	Coeff.	|t|	Coeff.	|t|
Numbers of other independent variables in the same equations								
More density variables	2		2		2		2	
Landscape variables	8		8		8		8	
Parent materials	9		9		9		9	
Soil classes	20		20		20		20	
Precipitation variables	2		2		2		2	
Others (texture, depth, source bias)	8		8		10		12	
Adjusted R-squared	.406		.472		.533		.495	
Dependent variable mean	0.8997		−2.1106		−3.2203		0.4521	
Standard error of estimate	0.5578		0.5350		0.6666		0.5594	
Least-squares type	GLS		GLS		GLS		GLS	
Number of observations, degree of freedom	1,223, 1,163		1,194, 1,134		1,182, 1,121		1,042, 980	
Observations from 1930s	32		9		22		22	
1950s	56		60		49		27	
1980s	1135		1125		1111		993	

Notes: See notes to table 4.1.

Table 4.3
Determinants of pH levels in north China, 1930s–1980s

Independent variables	Equation numbers and dependent variables							
	(1) Northwest		(2) The Huang-Huai-Hai plain		(3) Coast		(4) Northeast provinces	
	Coeff.	|t|	Coeff.	|t|	Coeff.	|t|	Coeff.	|t|
Time period and region (relative to 1980s, same region; − = positive growth up to 1980s)								
1930s, whole sample			−0.004	(0.04)			−0.767	(3.53)**
1930s, coast			0.974	(3.42)**	0.797	(1.99)[a]		
1930s, desert-steppe	0.410	(1.82)[a]						
1930s, spring wheat	−0.063	(0.64)						
1930s, winter wheat–millet	0.268	(2.90)**						
1930s, Huang-Huai-Hai					1.517	(2.82)**		
1950s, whole sample			0.129	(0.74)			0.408	(0.76)
1950s, desert-steppe	0.427	(1.88)[a]						
1950s, spring wheat	0.001	(0.02)						
1950s, winter wheat–millet	0.136	(0.70)						
Agricultural region	*(vs. winter wheat–millet)*				*(vs. north Chang Jiang)*			
Desert-steppe region	−0.148	(0.81)			−1.483	(2.56)*		
Spring-wheat region	0.244	(3.21)**			−1.305	(1.12)		
Huang-Huai-Hai plain					−0.109	(3.61)**		
South Chang Jiang plain					−1.355	(2.32)*		
Double-cropped rice								
Densities of human employment in each type of county in 1982								
ln (agricultural), T	0.008	(0.23)	0.051	(1.55)	−0.379	(0.53)	−0.008	(0.07)
ln (nonagricultural), T	0.010	(0.30)	−0.014	(0.40)	−0.273	(3.23)**	0.043	(0.42)

Table 4.3
(continued)

Independent variables	Equation numbers and dependent variables			
	(1) Northwest	(2) The Huang-Huai-Hai plain	(3) Coast	(4) Northeast provinces
Numbers of other independent variables in the same equations				
More density variables	2	2	3	3
Landscape variables	8	8	5	8
Parent materials	8	9	0	8
Soil classes	22	14	3	18
Precipitation variables	2	2	2	2
Others (texture, depth, etc.)	9	9	8	7
Adjusted R-squared	0.950	0.892	0.752	0.647
Dependent variable mean	8.497	7.861	7.427	7.061
Standard error of estimate	0.543	0.500	0.642	0.650
Least-squares type	GLS	GLS	OLS	OLS
Number of observations, degrees of freedom	682, 620	492, 442	126, 95	261, 210
Observations from 1930s	97	81	26	31
1950s	130	31	0	9
1980s	455	380	100	221

Notes: See notes to table 4.1.

Table 4.4
Determinants of soil pH levels and acidity in south China, 1930s–1980s

	Equation numbers and dependent variables							
	(1)–(2): Southwest, south and Sichuan				(3)–(4): East (Chang Jiang plain)			
	(1) pH level		(2) Acidity		(3) pH level		(4) Acidity	
Independent variables	Coeff.	|t|	Coeff.	|t|	Coeff.	|t|	Coeff.	|t|
Time period and region (relative to 1980s, same region; − = positive growth up to 1980s)								
1930s, coast					-0.0612	(0.30)	1.1466	(0.00)
1930s, Sichuan basin	0.1142	(0.63)	-0.3024	(1.49)				
1930s, north Chang Jiang					0.0749	(0.53)	-0.0878	(0.52)
1930s, south Chang Jiang					-0.3613	(2.79)**	0.2060	(1.70)[b]
1930s, double-crop rice	-0.7133	(3.77)**	0.9614	(6.14)**				
1930s, southwestern rice	-0.5079	(3.55)**	0.5587	(4.06)**				
1950s, Sichuan basin	-1.3319	(1.89)[a]	1.4910	(2.22)*				
1950s, north Chang Jiang					0.2718	(1.12)	-0.6590	(2.46)*
1950s, south Chang Jiang					0.6017	(2.55)*	-0.5743	(2.81)**
1950s, double-cropping rice	0.1061	(0.38)	-0.0545	(0.20)				
1950s, southwestern rice	0.7187	(3.16)**	-0.0911	(2.94)**				
Agricultural region	*(other than Sichuan basin)*				*(other than north Chang Jiang plain)*			
Desert-steppe region	0.2241	(0.67)	0.0401	(0.10)				
South Chang Jiang plain					-0.4085	(4.13)**	0.4032	(4.55)**
Double-cropping rice	-0.2113	(1.13)	0.0676	(0.34)				
Southwestern rice	-0.2182	(1.38)	0.1790	(1.03)				

Table 4.4
(continued)

	Equation numbers and dependent variables							
	(1)–(2): Southwest, south and Sichuan				(3)–(4): East (Chang Jiang plain)			
	(1) pH level		(2) Acidity		(3) pH level		(4) Acidity	
Independent variables	Coeff.	|t|	Coeff.	|t|	Coeff.	|t|	Coeff.	|t|
Densities of human employment in each type of county in 1982								
ln (agricultural), T	0.0745	(1.54)	−0.0089	(0.19)	−0.0695	(1.80)[a]	0.0266	(0.70)
ln (nonagricultural), T	−0.0018	(0.04)	−0.0861	(1.79)[b]	0.1034	(2.32)*	−0.0821	(1.98)[a]
Numbers of other independent variables in the same equations								
More density variables	2		2		2		2	
Landscape variables	8		8		8		8	
Parent materials	8		8		8		8	
Soil classes	22		22		14		14	
Precipitation variables	2		2		2		2	
Others (texture, etc.)	8		8		9		9	
Adjusted *R*-squared	.385		.370		.950		.545	
Dependent variable mean	5.8717		0.3481		6.8333		0.0695	

Table 4.4
(continued)

| | Equation numbers and dependent variables | | | |
| | (1)–(2): Southwest, south and Sichuan | | (3)–(4): East (Chang Jiang plain) | |
Independent variables	(1) pH level	(2) Acidity	(3) pH level	(4) Acidity
Standard error of estimate	0.9305	0.5284	0.7029	0.3401
Least-squares type	OLS	tobit	GLS	tobit
Number of observations, degree of freedom	860, 798	860, 798	727, 675	727, 675
Observations from 1930s	241	241	171	171
1950s	38	38	27	27
1980s	581	581	529	529

Notes: See notes to table 4.1.

for the desert-steppe region. Nor did the 1930s data show any hints of the post-1950s decline.

Unlike the results for organic matter and nitrogen, the total soil endowments of phosphorus and potassium definitely did not fall in the north in the period under study. Both phosphorus and potassium rose significantly from the 1930s to the 1980s in northern regions other than the Shaanxi-Shanxi region of winter wheat and millet. The rise looks more impressive in the earlier (1930s–1950s) transition, to judge from table 4.5 and figure 4.4. In terms of statistical significance, however, the clearest rise occurred after the 1950s in the same relatively well-sampled Nei Mongol area where organic matter and nitrogen levels declined. We return to the task of interpreting this rise in phosphorus and potassium after introducing a similar result for south China.

The Threat of Alkalinity and Salinity, and Local Victories

Another important soil chemical trend to follow in the north relates to salinity, sodicity, and alkalinity. Extreme degrees of these interrelated conditions can be toxic for crops, and milder degrees of each can restrict crop choice.[7] It has been feared since at least the 1930s that increasingly saline and alkaline soils in the northwest and on parts of the coast would reduce the area available for staple food crops.

Although the data do not permit direct measures of trends in salinity, they do permit us to comment on trends in alkalinity, a related concern in the same arid and coastal areas of north China that are vulnerable to salinity. Thus we use topsoil alkalinity (high pH) as an indirect clue, a spatial correlate, of serious sodicity and salinity in the root zone (see Lal, Hall, and Miller 1989, pp. 64–65; Institute of Soil Science 1990, chap. 15). The trends in alkalinity are, in turn, shown indirectly by the regression-based pH results in table 4.5 and figure 4.4. For simplicity of exposition, we show only the pH results here,[8] given that all key results are the same for pH as for alkalinity defined as $Alk = \max (pH - 8, 0)$.

No trend toward alkalinity shows up in the pH regression results, either for major regions or for well-sampled subregions. The closest thing to rising alkalinity is the statistically insignificant slight rise in pH for the spring wheat region from the 1930s to the 1950s. No other rise in pH occurs in an area where alkalinity prevailed. On the contrary, alkalinity

Table 4.5
Estimated average levels of soil nutrients and pH on tilled fields in north China, 1930s–1980s

Agricultural region	Organic matter (%)			Nitrogen (%)		
	1930s (n)	1950s (n)	1980s (n)	1930s (n)	1950s (n)	1980s (n)
All north (except desert-steppe)	1.48 (24)	1.43 (124)[a]	1.13 (989)	0.080 (21)	0.088 (117)	0.073 (972)
Spring wheat region	— (5)	2.08 (87)[a]	1.52 (431)	— (3)	0.102 (96)	0.096 (416)
Winter wheat–millet region	1.30 (15)	1.73 (9)	1.25 (162)	0.067 (15)	0.081 (11)	0.077 (163)
Huang-Huai-Hai plain	— (4)	1.10 (28)	0.94 (396)	— (3)	0.091 (10)*	0.062 (393)

Agricultural region	Phosphorus (%)			Potassium (%)		
	1930s (n)	1950s (n)	1980s (n)	1930s (n)	1950s (n)	1980s (n)
All north (except desert-steppe)	0.043 (24)	0.050 (28)	0.048 (959)	1.26 (25)**	1.73 (24)	1.90 (696)
Spring wheat region	— (2)	0.027 (17)*	0.044 (413)	— (2)	1.28 (14)*	1.92 (368)
Winter wheat–millet region	0.071 (17)	— (5)	0.057 (163)	1.61 (18)	— (5)	1.75 (110)
Huang-Haui-Hai plain	— (5)	0.061 (6)	0.047 (383)	—	— (5)	— (5)

Agricultural region	pH		
	1930s (n)	1950s (n)	1980s (n)
All north (except desert-steppe)	7.52 (200)[b]	7.76 (148)	7.66 (1045)
Spring wheat, northeast	6.53 (31)**	7.28 (9)	7.13 (221)
Spring wheat, northwest	8.16 (34)	8.30 (96)	8.35 (288)
Western Gansu	8.46 (26)	8.06 (10)	8.38 (32)**
Huang He Great Bend	— (0)	8.92 (32)	8.53 (56)
Chifeng municipality	— (0)	8.70 (17)	8.63 (50)

Table 4.5
(continued)

Agricultural region	Organic matter (%)		
	1930s (n)	1950s (n)	1980s (n)
	pH		
Winter wheat–millet region	8.13 (54)**	8.03 (12)	7.82 (156)
Wei Valley, Shaanxi	7.36 (26)	— (2)	6.83 (24)
Huang-Huai-Hai plain	7.44 (81)	7.65 (31)	7.53 (380)
Bohai and North Jiangsu coast	9.20 (26)**	— (0)	6.88 (100)

Notes: — denotes an average that is not displayed here, or in figures 4.4 or 4.5, because it is based on fewer than six representative profiles for the 1930s or 1950s.

Averages are based on the regressions in tables 4.1–4.4.

Averages for the 1930s and 1950s represent the same mixtures of site attributes as those for actual tilled fields in the 1980s. Specifically, each average for the 1930s or 1950s is a noisy back-cast equal to the actual 1980s figure for tilled fields minus the sum of the regression-predicted historical shift and the average change in prediction error from all the earlier periods' profiles to all those of the 1980s.

*n*s represent the total number of profiles in a given era and region, not just those tilled.

All-north and all-south averages are derived from the regional averages using the number of tilled plots in the 1980s as regional weights. Averaging is geometric (logarithmic) for nutrients, but not for pH.

Significance of the difference between a 1930s or 1950s average and its 1980s counterpart: **significant at the 1-percent level (two-tailed), *significant at 5-percent level, [a]significant at 7-percent level, [b]significant at 10-percent level.

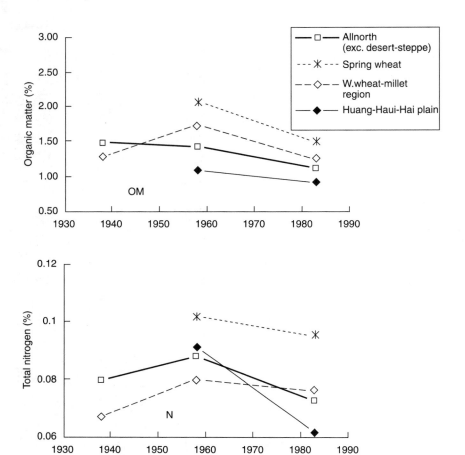

Figure 4.4
Estimated average levels of soil nutrients and pH on tilled fields in north China, 1930s–1980s

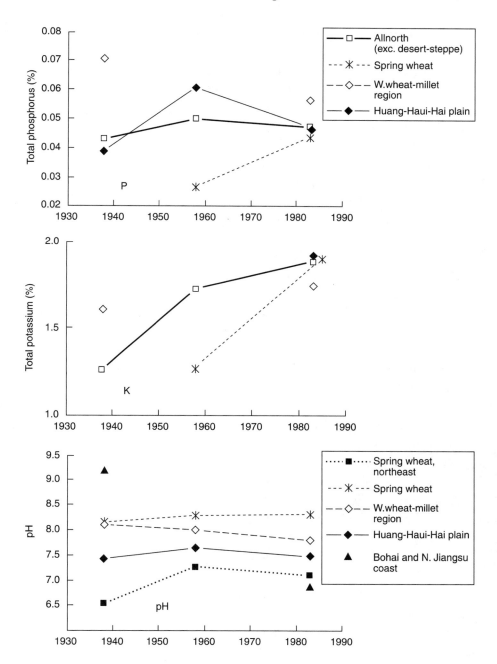

seems to have dropped dramatically along the Bohai and north Jiangsu coasts from the 1930s to the 1980s, an occurrence we return to shortly. Over that same half century, alkalinity also abated in the Wei Valley and the rest of the Shaanxi-Shanxi area, as suggested by table 4.5's regression results for the winter wheat–millet region.

These suggestions from pH regressions can be cross-checked and extended by taking advantage of a pair of independent sources from 1936 and 1986. In 1936, two pioneers of sampling and mapping China's soils ventured a guess about the patterns of average pH levels all over China. Although their guesses were only meant to be tentative, they are worth examining. Figure 4.5 gives the overall pH contours envisioned by the young Chinese soil scientist Li Chingkwei and his American mentor James Thorp. If we think of serious alkalinity as any average pH over 8.0, then Li and Thorp believed alkalinity to be serious in the desert northwest; in the whole valleys of the Huang (Yellow), Huai, and Hai rivers from Shaanxi to the sea; and in the lowland center of the northeast provinces. If we limit our view to the shaded areas where pH averaged above 8.5, the area of severe alkalinity consists of a few desert patches, the great bend of the Huang He, the Bohai and Jiangsu coasts, and again the central lowlands of the northeast.

The two portrayals of pH in the north in the 1930s—the regression-based estimates and the map in figure 4.5—are similar. This is not surprising, since Li and Thorp and their coworkers were responsible for both the soil profile numbers and the map. There is a discrepancy between the two, however. The full version of the soil profile regressions makes use of a mixture of pH data, some from profiles and some from the figure 4.5 map itself. When the two are mixed together in statistical pH regressions, the map-based pH figures appear to understate the pH in the Huang-Huai-Hai plain by 0.46, though no such significant discrepancy is found for the map's figures for the area along the coasts.[9] The tentative interpretation offered here is that alkalinity was more serious in the Huang-Huai-Hai plain of Henan, Hebei, northern Anhui, and inland Jiangsu back in the 1930s than figure 4.5's map suggests.

For the 1980s we have the benefit of a corresponding pH map published by the Institute of Soil Science, Academia Sinica, in 1986. Drawing on early returns from the second National Soil Survey for the early 1980s, the institute produced the pH map that is summarized in figure 4.6. The map and the soil profile numbers agree with respect to pH in the north, at least for the parts of the north covered in the soil profile data for 1981–1986.

Alkalinity in the 1890s was most severe in figure 4.6's shaded areas representing an average pH above 8.5. Two of these areas, the great bend area of the Huang He and the central lowlands of western Jilin and Heilongjiang in the northeast, match areas of severe alkalinity in the Li and Thorp sketch from half a century earlier. Beyond these two areas, however, the two pictures diverge. By the 1980s, say the maps, alkalinity may have become more severe in the desert areas of Gansu and western Nei Mongol, to the west of the Great Bend area. On this western desert fringe the pH trend must remain somewhat in doubt, since neither the 1936 map nor the 1930s soil profile data offered much information from truly desert areas.[10]

On the other hand, comparing the pH maps from 1936 and 1986 confirms that pH improved in important agricultural areas along the coast. Note that in figure 4.6 the Bohai coast and the Jiangsu coast no longer show up as areas of severe alkalinity. On these local victories the map evidence of 1936 and 1986 agrees with the regression results based on soil profiles.

The victories over alkalinity and salinity in these two coastal areas were part of the larger struggle for water control throughout the Huang-Huai-Hai plain and its coasts in the time between the two maps. In the upper reaches of the Huang, Huai, and Hai Rivers, the emphasis was on flood control and irrigation. The lower reaches close to the coast required additional investments in drains, pumps, and polder fields to minimize waterlogging and flush out salts. Considerable attention was also devoted to changing the seasonal pattern of water application and drainage to lower the water table rapidly at certain times of the year, subject to the constraint that groundwater reserves not be depleted. At the coastline itself, new tidal barriers, escape canals, and pumps translated the plain's fight against riverine waterlogging into a fight against seawater ingression (Institute of Soil Science 1990, chaps. 38, 39).

The progress of the Huang-Huai-Hai investments in water control did not proceed evenly. Little was accomplished in the 1930s and 1940s

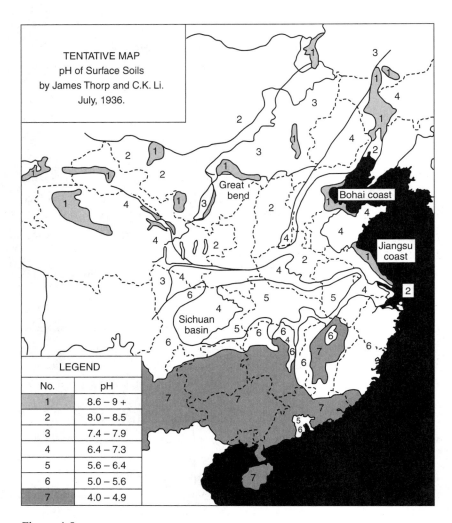

Figure 4.5
Li-Thorp sketch of average pH ranges for China, 1936
Source: Thorp 1939.

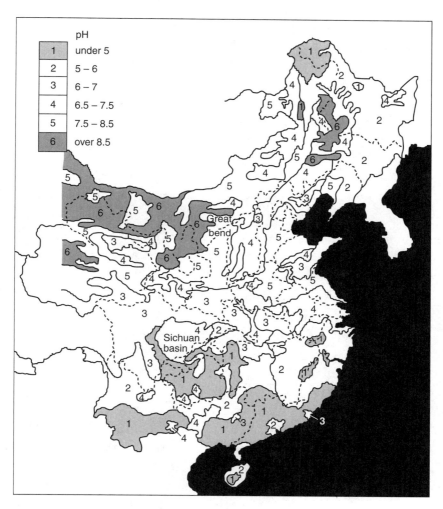

Figure 4.6
Soil Atlas sketch of average pH ranges in China, 1986
Source: Based on Institute for Soil Science 1986, pp. 33–34.
Note: See also Institute for Soil Science 1990, fig. 27.12.

because of economic depression and warfare. The main investments in harnessing the three rivers and improving drainage came in the first decade after the revolution of 1949. However, the drought conditions of 1958–1961 combined with some excesses of the Great Leap to bring overirrigation of the plain, raising the water table and accelerating secondary salinity. After 1962, however, a renewed phase of reduced irrigation and greater attention to drainage reduced salinity and, presumably, alkalinization in the lowlands once again (Institute of Soil Science 1990, pp. 727–28).

For the north as a whole, then, the best reading of the pH trends is mixed:

• Continued severe alkalinity, with no trend, in the great bend of the Huang He and in the lowlands of the northeastern provinces of Jilin and Heilongjiang.

• A possible worsening of alkalinity (and salinity) in the desert areas of the far northwest.

• Local victories over alkalinity and salinity in two agriculturally important coastal areas.

Soil Trends in South China

pH and Acidity

Comparing regression evidence with hints from the pH maps of 1936 and 1986 also illuminates the history of the acidity problem in the tilled soils of south China. The trends in acidity implied by soil profile data are represented by the trends in pH in table 4.6 and figure 4.7.[11] By either the pH or the acidity yardstick, the prevailing changes resulted in a clear net improvement—a reduction in acidity—from the 1930s to the 1950s, followed by a partial reversion toward greater acidity from the 1950s to the 1980s. The 1930s–1950s improvement was strongly significant in every agricultural region south of the Chang Jiang (Yangzi). From the 1950s to the 1980s the same large area saw a drift back toward acidity that was neither zero nor large enough to prevent a net improvement from the 1930s to the 1980s.

The changing geography of soil acidity over a whole half century can be verified by again comparing the soil profile evidence with the 1936 and 1986 maps in figures 4.5 and 4.6. If acidity is defined as an average pH below 6, the 1936 map implies that acidity was serious everywhere below (roughly) the Chang Jiang. On the narrower definition of pH below 5, the shaded area in figure 4.5 still covers most of Yunnan, Guizhou, Guangxi, Guangdong, Hunan, and Jiangxi. The regression evidence and the map agree in most respects. Like the regression estimates, both maps put the average pH below 5 for the double-cropping region extending from eastern Guangxi through southern Fujian, with the shared exception of the intensely farmed Zhu Jiang (Pearl River) area around Guangzhou (Canton). Both also show pockets of high acidity involving Jiangxi, though the pockets seem to have moved between 1936 and 1986.

For the southwestern rice region (Yunnan, Guizhou, and western Guangxi), however, the averages in table 4.6 probably give a truer picture of the net change in pH than does a comparison of the 1936 and 1986 maps. The soil profile evidence for this region casts doubt on the broad-brush picture painted by the tentative map of 1936. The statistical procedure of supplementing missing pH data with readings from the map suggests that the map has understated pH by 0.56 for this region. If so, then the true pH pattern in the southwest was probably about the same in 1936 as it appears in that part of the 1986 map shown in figure 4.6. The best guess is the acidity of the southwestern rice region did not change between the 1930s and 1980s.

It is not surprising that soil acidity continued to follow the contours of the rice-growing region during this period, dominating the double-cropping rice region in particular and also persisting in the southwest rice region and parts of Jiangxi. Rice is more acid-tolerant than most field crops, including wheat and many vegetable crops. Dealing with acidic soils in the subtropics means preferring rice (and tree crops) over other crops. If market conditions called for a switch to less acid-tolerant field crops, the conversion would take several years of liming to raise pH, especially on clay soils like those in the southwest. Chapters 5 and 6 will shed some indirect light on the possible economic cost of the restrictions soil acidity places on crop choice.

Table 4.6
Estimated average levels of soil nutrients and pH on tilled fields in South China, 1930s–1980s

Agricultural region	Organic matter (%)			Nitrogen (%)		
	1930s (n)	1950s (n)	1980s (n)	1930s (n)	1950s (n)	1980s (n)
All south China	1.60 (32)[b]	2.23 (56)[a]	2.37 (1129)	0.128 (9)	0.164 (60)	0.128 (1120)
Sichuan basin	1.32 (15)	— (2)	1.78 (118)	— (2)	— (4)	0.109 (118)
North Chang Jiang plain	2.53 (7)	2.21 (13)	2.10 (212)	0.153 (7)	0.183 (15)[b]	0.129 (209)
South Chang Jiang plain	1.27 (7)	2.59 (11)	2.51 (310)	— (0)	0.187 (10)[b]	0.140 (304)
Double-crop rice (far south)	— (0)	2.17 (13)	2.09 (270)	— (0)	0.118 (15)[b]	0.107 (270)
Southwestern rice	— (3)	2.11 (17)**	3.34 (219)	— (0)	0.164 (16)	0.189 (219)

Agricultural region	Phosphorus (%)			Potassium (%)		
	1930s (n)	1950s (n)	1980s (n)	1930s (n)	1950s (n)	1980s (n)
All south China	0.039 (22)	0.042 (49)	0.048 (1106)	0.730 (21)**	1.303 (27)*	1.430 (988)
Sichuan basin	0.053 (6)	— (4)	0.045 (115)	0.330 (6)**	— (4)	1.470 (102)
North Chang Jiang plain	0.047 (7)	0.058 (9)	0.057 (200)	1.507 (6)	— (2)	1.800 (167)
South Chang Jiang plain	0.032 (6)	0.037 (6)	0.042 (305)	0.450 (6)**	— (1)	1.642 (278)
Double-crop rice (far south)	— (0)	0.040 (14)	0.046 (266)	— (0)	0.759 (12)	1.036 (260)
Southwestern rice	— (3)	0.071 (16)	0.058 (220)	— (3)	0.412 (8)*	1.068 (181)

Table 4.6
(continued)

Agricultural region	Organic matter (%)			Nitrogen (%)		
	1930s (n)	1950s (n)	1980s (n)	1930s (n)	1950s (n)	1980s (n)
	pH					
Sichuan basin	6.64 (90)	— (2)	6.52 (68)			
North Chang Jiang plain	6.94 (113)	7.17 (15)	6.92 (211)			
South Chang Jiang plain	5.50 (57)**	6.42 (12)**	5.78 (318)			
Double-crop rice (far south)	5.21 (57)**	6.03 (16)	5.92 (270)			
Southwestern rice	5.70 (94)**	6.93 (20)**	6.21 (227)			

Notes: See notes to table 4.5.

Significance of the difference between a 1930s or 1950s average and its 1980s counterpart: **significant at the 1-percent level (two-tailed), *significant at 5-percent level, [a]significant at 7-percent level, [b]significant at 10-percent level.

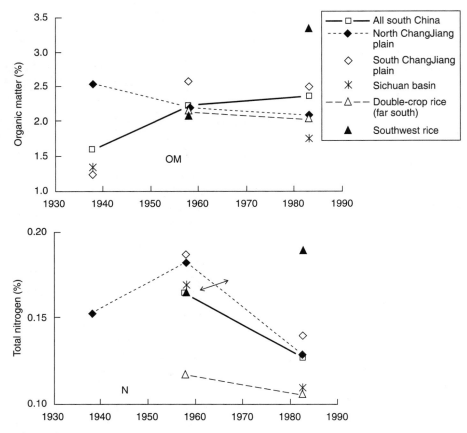

Figure 4.7
Estimated average levels of soil nutrients and pH on tilled fields in south China,
1930–1980s

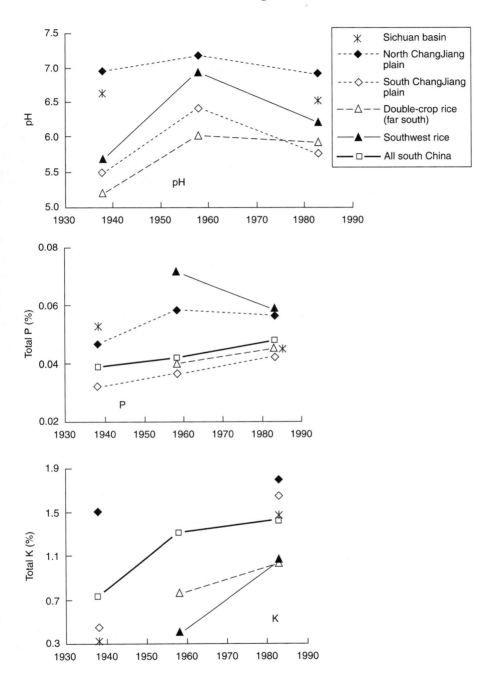

Organic Matter and Nitrogen

South China's trends in organic matter and in nitrogen can be illuminated in detail only for the period from the 1950s to the 1980s, with just a few hints about the 1930s.[12]

Trends in soil organic matter between the 1950s and the 1980s show a striking contrast between the southwest rice region and the rest of southern China. Most regions experienced no change or a statistically insignificant decline. Yet the organic-matter content of southwestern soils rose strongly, according to the figures. To follow the trends back to the 1930s, we must be content with small-sample hints from just one region: In the wheat-rice region north of the Chang Jiang, the stagnation or slight decline in organic matter might have been going on since the 1930s or earlier, according to table 4.6 and figure 4.7.

The trends in total nitrogen content from the 1950s to the 1980s show the same contrast between the southwestern rice region and the rest of south China. There is again a large estimated rise in the southwest, though without clear statistical significance. In other parts of the south, the estimated decline is greater for nitrogen than for organic matter, with only marginal statistical significance. Nothing can be said about trends in nitrogen content between the 1930s and later decades, since we have nitrogen data from the 1930s for only nine cases, seven of which are from the north Chang Jiang plain.[13] For the present, then, our tentative conclusions about trends in nitrogen are confined to the 1950s–1980s change. Total soil nitrogen fell in this interval over most of south China but rose in the southwest rice area.

The Phosphorus-Potassium Buildup

Trends look quite different for the phosphorus and potassium contents of south China's soils. Here every hint points to an upward trend, except for the estimated decline in the southwestern rice region's phosphorus content after the 1950s. For potassium, the upward trend looks monotonic and significant over the whole half century (1930s–1980s), just as it does for northern China. For phosphorus, the rise is less obvious, but still likely.

Could the trend in total potassium have been as steep as table 4.6 and figure 4.7 imply? Since the 1980s values agree with those given by the

Institute of Soil Science (1986, 1990), any suspicions should focus on the very low total potassium values for the 1930s and 1950s. One possible source of error is the likelihood that some of the 1930s and 1950s sources may have reported K values as if they wee K_2O values, causing a 20-percent understatement when the K_2O values were adjusted to yield K equivalents. Yet even if all data had been subject to such an error, table 4.6 suggests that a full 20-percent adjustment would still leave a steep trend. Another possibility, left unresolved here, is that methods of potassium extraction changed during the period in a way not yet revealed. Again, however, it is hard to see how changes in testing methods could have created so steep an artificial trend.

Our tentative conclusion for the south as a whole, taking account of the smallness of the 1930s samples and the range of statistical error, is that the levels of total potassium and total phosphorus rose between the 1930s and 1950s and again from the 1950s to the 1980s. Quantifying that net rise, however, would require larger samples for the 1930s and 1950s.

Interpreting the Trends in Soil Nutrients

Three Kinds of Patterns to Reconcile

What forces could explain the main patterns in China's soil chemistry since the 1930s? Until the data set is expanded, we should venture only a few broad interpretations here and concentrate on interpreting the patterns in soil nutrient levels, setting aside the trends in pH. Specifically, three soil nutrient patterns need to be interpreted:

• *The temporal pattern.* Why were the trends in nutrient levels more favorable from the 1930s through the 1950s and less favorable from the 1950s to the early 1980s?[14]

• *The east-west pattern for organic matter and nitrogen.* Why did levels of organic matter and nitrogen seem to stagnate or drop in the heavily populated east and rise only in some western regions? In the south, for example, why did levels of organic matter and nitrogen stagnate or even decline from the 1950s to the 1980s in most of south China, yet rise impressively in southwest China? Similarly, in the north, why did levels of organic matter and nitrogen seem to drop at least as much across the

Huang-Huai-Hai plain and in the spring wheat region as in the Shaanxi-Shanxi winter wheat zone?

• *The chemical pattern.* Why were there signs of decline in soil organic matter and nitrogen, but not in phosphorus and potassium? Other clues seem to confirm that both the trend and level of soil nitrogen were more serious than those for phosphorus or potassium. The share of soils sampled in China's soil test network that were nitrogen-deficient rose from about 74 percent in 1935–1941 to about 100 percent in the 1970s and early 1980s. Although reports of phosphate and potash deficiencies also rose from 1935–1941 to the 1980s, they rose more slowly, and remained less frequent, than the nitrogen deficiencies throughout this half century (Stone 1988, pp. 812–18). Why?

I offer only some tentative partial interpretations here, in the hope of stimulating further work on these issues.

What Role for Erosion?

To address the prevailing suspicion that erosion is the main source of soil degradation, we must be content with indirect clues. One indirect physical clue, topsoil layer thickness, does suggest an erosion trend, at least at first glance. Yet as we shall see, that clue may be a mirage, and trends in soil chemistry lend less support to the belief that erosion is the dominant form of soil degradation in China.

Other things being equal, topsoil losses should cause a thinning of what soil scientists call the "A horizon," the first soil "horizon" (soil layer with relatively uniform characteristics) below surface debris. Table 4.7 reveals what has happened to the topsoil depth in China, first in the history of some simple sample averages, and then in regression results.

At first glance, the trends in A-horizon thickness may seem to imply serious topsoil loss between the 1930s and the 1950s, but no further losses from the 1950s to the 1980s in most regions (table 4.7, panel A). This change of trend would conflict with the common belief that erosion is accelerating. It also would clash with the belief that human cultivation is a leading contributor to topsoil losses on already tilled lands, since the losses occurred in the 1930s–1950s period, when the intensity of cultivation was falling, not during the later rise in cultivation intensity. The

A-horizon measures thus suggest that erosion may have occurred, but not so recently as has been presumed.

The top-layer thickness measures we have are not comparable over time, however. It is not clear that they are really A horizons according to the definition that some soil scientists consider relevant in terms of measuring topsoil loss. Our data sources for the 1930s and 1950s usually did not give their soil layers letter names. Perhaps what looks like a thinning of the A horizon between the 1930s and the 1950s is really a switch from broad and loose definition to a tighter definition of what constitutes a meaningful horizon. Perhaps the later stability also masks shifts in the definition of a soil layer. We know that by the 1980s a bare majority of profiles referred to "A horizons," whereas others either gave no labels or adopted narrower definitions of that first major horizon (A_p, A_{11}, "tilled layer," etc.)

The middle column of numbers in panel A of table 4.7 offers a further clue to the importance of shifts in definition. These are the shares of all soil profiles whose recorders decided to round off the bottom depth of the topsoil layer in the profile to a number of centimeters ending in zero or five. When the share of these depths ending in zero of five far exceeds the random-chance 20 percent, we know that the recorders did not bother to define a horizon carefully. As can be seen, that share of rounding was a high 58.1 percent in the 1930s, but only 43.8 percent in the 1950s and only 36.5 in the 1980s. Apparently, the 1930s measurements involved a less careful definition of a soil horizon than the later measures. This apparent indifference to precision may well have been more characteristic of deeper samples taken for geological rather than agronomic purposes. If so, then the "thinning" of the topsoil after the 1930s may be only an artifact of getting more precise abut defining horizons.

To support this conjecture, consider the two sets of regression results in Panel B of the same table, which show how the three eras (1932–1944, 1950–1964, and 1981–1986) differed in their topsoil depths, holding other things constant. The first set of coefficients, labeled "without correction," simply uses the same control variables we used to reveal trends in the chemical characteristics. These seem to imply that between the 1930s and the 1950s most of those regional topsoil losses were

Table 4.7
Topsoil A-horizon thickness in China, 1930s–1980s

Panel A. Trends in average thickness of the topsoil layer, and in statistical "0–5 rounding"

Era	Average thickness of topsoil layer (in centimeters)	"0–5 rounding": percentage of profiles with topsoil layer's bottom depth recorded as ending in 0 or 5 cm	Number of soil profiles
Northwest China			
1930s (1932–1944)	23.2	64.9	97
1950s (1950–1964)	15.1	43.1	130
1980s (1981–1986)	20.4	40.4	449
Huang-Huai-Hai plain			
1930s (1932–1944)	28.6	65.4	78
1950s (1950–1964)	14.3	41.9	31
1980s (1981–1986)	18.9	51.1	380
Northeast provinces			
1930s (1932–1944)	23.3	93.5	31
1950s (1950–1964)	11.9	88.9	9
1980s (1981–1986)	19.5	47.0	219
East China (Chang Jiang plain)			
1930s (1932–1944)	24.4	46.8	171
1950s (1950–1964)	17.0	37.0	27
1980s (1981–1986)	15.5	29.3	529
Southwest China			
1930s (1932–1944)	19.9	56.4	241
1950s (1950–1964)	16.8	42.1	38
1980s (1981–1986)	16.0	26.7	581

Table 4.7
(continued)

Era	Average thickness of topsoil layer (in centimeters)	"0–5 rounding": percentage of profiles with topsoil layer's bottom depth recorded as ending in 0 or 5 cm	Number of soil profiles
All China			
1930s (1932–1944)	22.9	58.1	618
1950s (1950–1964)	15.4	43.8	235
1980s (1981–1986)	17.6	36.5	2158

Panel B. Regression-revealed shifts in A horizon thickness, with and without correction for the "0–5" rounding

	Without correction				With correction for rounding			
	1930s → 1980s		1950s → 1980s		1930s → 1980s		1950s → 1980s	
	Coeff.	|t|	Coeff.	|t|	Coeff.	|t|	Coeff.	|t|
NW spring wheat area	1.19	(0.64)	**8.03**	(6.23)	-2.71	(1.85)	**2.24**	(2.46)
Winter wheat–millet	-5.33	(2.85)	-1.19	(0.39)	-1.99	(1.43)	0.14	(0.07)
Huang-Huai-Hai plain	**-9.96**	(8.25)	2.40	(1.31)	**-6.48**	(7.32)	-0.62	(0.54)
Northeast	-4.30	(1.69)	7.65	(1.22)	-0.74	(0.43)	-0.29	(0.07)
North Chang Jiang plain	**-6.12**	(5.01)	-1.39	(0.65)	**-4.42**	(5.50)	-0.37	(0.29)
South Chang Jiang plain	**-7.61**	(6.50)	-3.01	(1.40)	**-2.91**	(4.04)	0.45	(0.35)
Sichuan basin	-4.79	(3.14)	8.29	(1.38)	-2.75	(2.79)	-2.49	(0.64)
Double-crop rice area	-4.82	(3.34)	-0.51	(0.21)	-2.34	(2.43)	0.74	(0.48)
Southwestern rice area	-0.47	(0.39)	-1.41	(0.72)	-0.23	(0.30)	**-3.17**	(2.54)

Table 4.7
(continued)

Notes: Data are from the China soil history samples described in chapter 3.
Agricultural regions are those mapped in figure 4.3.

Northwest China here consists of the spring wheat and winter wheat–millet areas excluding the northeast provinces, which are given separately. "Southwest China" consists of the Sichuan basin, the southwestern rice area and the double-cropped rice area.

The underlying regressions are the same as the pH regressions in appendix table A.4, except that (a) the dependent variable is the thickness of the topsoil layer currently designated as the A horizon, in centimeters, and (b) the regressions that include correction for rounding add the BOTTOM05 independent variable defined as follows:

BOTTOM05 = 1 if the depth of the bottom of the topsoil layer is a number of centimeters ending in 0 or 5.
BOTTOM05 = 0, otherwise.

Thus BOTTOM05 serves as an indicator that the experimenters rounded off the bottom depth because its exact magnitude was of little concern. The lack of concern over precision is taken as an indication of low interest in defining a true horizon with consistent characteristics. For the same discussion of rounding and topsoil depths in Indonesia, see the section in chapter 7 on erosion, particularly the text surrounding table 7.5.

Regional statistical samples are the same as those used to estimate pH.

statistically significant. These coefficients might appear to confirm the prevalence of serious erosion for the period from the 1930s to the 1950s.

Yet we must consider the role of sampling motivation, as evidenced in statistical rounding, as the right-hand set of regression estimates in Panel B, labeled "with correction for rounding," does. These estimates show that when one has corrected for whether or not the bottom depths of the samples were rounded off, the thinning looks less impressive even between the 1930s and the 1950s, and between the 1950s and the 1980s, the figures offer no evidence of significant thinning anywhere but in the southwestern rice area. Note that in the northwest spring wheat area, where desertification was alleged to be causing great topsoil losses, the estimates corrected for rounding actually imply a *thickening* of the top-soil horizon from the 1950s to the 1980s.

Turning from physical to chemical clues, the three patterns in soil chemistry trends summarized above raise questions about erosion trends and the role of tillage in causing them. An extensive literature has noted that typical erosion on topsoil endowments seems to deplete *all* the macronutrients and organic matter (Follett and Stewart 1985, pp. 339–41; Larson, Pierce, and Dowdy 1983; El-Swaify, Moldenhauer, and Lo 1985, chaps. 22, 23, 28, 29). If topsoil erosion by wind and water were the dominant form of topsoil degradation, one would expect to see a downward trend in total nitrogen and organic matter *and* phosphorus *and* potassium whenever and wherever human-induced erosion is greater. But this is not what China's historical data show us.

First, the temporal pattern—less favorable trends in total organic mat-ter, nitrogen, phosphorus, and potassium after the 1950s than before—casts some doubt on the belief that the rate of erosion on China's tilled lands has risen cumulatively and monotonically with the intensity of cultivation. It is true, as we shall see, that one dimension of intensity of cultivation, the multiple-cropping index, fell from the 1930s to the 1950s then rose thereafter, suggesting that the earlier period might have experi-enced less erosion and therefore less nutrient depletion per year. But by all other measures the intensity of human and man-made inputs rose in both periods, so that cumulative erosion should have advanced in both peri-ods. Correspondingly, topsoil nutrients should have been depleted in both periods, if we are to use such simple clues to show that erosion was the dominant process of soil degradation. That was not the observed temporal pattern, however.

Second, the usual emphasis on erosion does not fit the spatial pattern in China very well. In north China, erosion was supposed to have been worst in the loess plateau erosion region, roughly the winter wheat–millet region (see figure 4.3), yet this region showed less sign of a decline in organic matter, total nitrogen, and total potassium than any other region. In south China, all erosion predictions imply that the loss of organic matter and nitrogen should have been at least as severe in southwest China as in the rest of south China, yet this runs contrary to the observed spatial pattern.

Third, the chemical pattern in nutrient trends—falling organic matter and total nitrogen but rising phosphorus and potassium—offers equivocal evidence on erosion trends. Some studies have found that erosion removes *available* phosphorus and potassium in even greater percentages than organic matter or nitrogen, yet we see that *total* phosphorus and potassium apparently rose over the period. Human-induced erosion might have left such a fingerprint, if topsoil removals brought more abundant, but less available, phosphorus and potassium closer to the surface. Whether that happened in China is a subject for further study.

Erosion, in other words, has not left its characteristic fingerprints on the broad trends in China's cultivated topsoils. The main suspicions, those imagining serious erosion on China's vast cultivated areas, fail to find confirmation in the trends in topsoil chemistry for China's cultivated lands. (Of course, erosion has continued on some uncultivated sites and on recently deforested areas such as the upper reaches of the Chang Jiang.)

Systematic Biases?
Another possible interpretation to consider is that the time trends may be misleading. Perhaps subtle biases changed the character of the soil profile samples between 1932–1944 and 1950–1964, and again between the latter period and 1981–1986. Suppose, for example, that the 1950–1964 data were relatively optimistic because those soil profiles sampled better soils than the profiles in the comprehensive survey of the 1980s or in the (pessimistic?) samplings of the 1930s. In particular, as we said above, the

profiles from 1958–1961 might have been chosen to show the best potential for raising yields during the Great Leap. Such shifts in selectivity could distort our comparisons over time.

Several devices have been used to detect and quantify such source biases. The technical details of these side tests for bias are discussed elsewhere.[15] Aside from these few clues, we have found no clear evidence that shifting biases in source materials have distorted the trends sketched in tables 4.5 and 4.6 and in figures 4.4–4.7. And the minor cases of possible bias tend more to reinforce than to undermine this chapter's broad conclusions about trends. With the few caveats just noted, the trends appear to be real.

Fertilizer versus Uptake

Toward Historical Nutrient Balances   The eventual explanation of the observed trends in soil macronutrients and organic matter will have to rest on historical nutrient balance accounts for each nutrient, breaking each historical net change in the topsoil endowment into its component inflows and outflows. Some fuller future investigations may reveal rough balances for several benchmark dates, such as 1933, 1957, 1983, 1990, 2000, and so on.

For any one nutrient the soil balance accounts are as complex as the underlying chemistry and physics. To see the overall contours of this historical accounting task, let us consider the forces shaping the gains, transformation, and losses of soil nitrogen. Similar forces apply to other macronutrients, namely, phosphorus, potassium, or sulfur (Institute for Soil Science 1990, chaps. 29–31; Brady 1990, chaps. 11–12), yet we focus on soil nitrogen here to underline the distinctiveness of its historical trends. Figure 4.8 shows the main sources and destinations of soil nitrogen, noting separately those governed mainly by changes in human interventions over the course of agricultural history, which are our main empirical targets. What do we know about the changes over time in human application of organic fertilizers, of commercial (synthetic chemical) fertilizers, in the uptake of nitrogen by crops, and in the losses to erosion? Having previously examined trends in erosion, we turn to the history of fertilizers and of crop uptake.

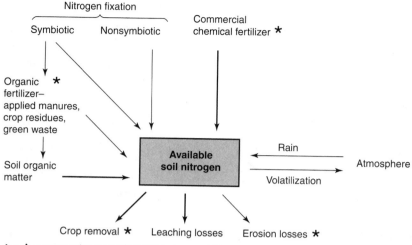

Figure 4.8
Major types of gains and losses of available soil nitrogen
Source: Adapted from Brady 1990, p. 337.
Note: Bolder arrows (from the original) represent the greater volumes of additions and losses.

Here we can make only a few initial steps toward such an accounting, concentrating on broad indirect historical measures of the inputs and demands on the soil that humans are most likely to have changed over time. The best tentative conjecture is that the observed patterns in soil nutrients look like responses to differences in agricultural uptake, differences that exceeded the differences in fertilizer application.

Fertilizer Patterns   Knowing the trends in fertilizer application would not in itself help us explain the three main patterns we have discerned in China's macronutrient history (temporal, spatial, and chemical). On the contrary, what we know about trends in commercial fertilizers actually *adds* to the mystery on all three fronts.

The temporal pattern identified was that the trends in soil endowments of nitrogen, phosphorus, and potassium generally showed fewer increases (or more decreases) in the later period from the 1950s to the 1980s than in the earlier 1930s–1950s period. For fertilizer to play the lead role in this pattern, fertilizer application would have had to drop, or at least greatly decelerate, after the 1950s. That was not the case, however, as these estimates on chemical fertilizer production[16] (in metric tons per year) show:

	All	Nitrogen	Phosphate	Potash
1957–1959	204	148	51	4
1982–1984	13,796	11,191	2,574	31

Virtually all of the increase in such manufactured fertilizers came after the late 1950s, both in the relevant absolute measures and even in percentage-growth terms. Other things equal, this acceleration should have caused a greater buildup of nitrogen, phosphorus, and potassium in the soil after the 1950s than before. Yet we have seen that the actual pattern was the opposite.

The spatial pattern is also the opposite of what one would expect from the data on fertilizer application. As shown in table 4.8 and figure 4.9, fertilizer use per hectare was always greater in the east than in the west, the opposite of what would be expected from the observed spatial pattern in soil nutrient levels and trends.

As for the chemical pattern, since 1959 use of nitrogen has shown the fastest rate of growth in chemical fertilizer application in China, potash the slowest, and phosphate in between, as we have just seen. These trend differences do not fit the pattern of soil endowment growth (K growth $\geq$ P growth $> 0 >$ N growth) that we have observed for soils over the 1950s–1980s period. Finally, nitrogen fertilizer application accelerated mostly after the 1950s, the period of stagnant or dropping levels of nitrogen.[17]

Agricultural Uptake   If the observed soil endowment patterns do not fit the erosion hypothesis or the differences in fertilizer application very well, then what other forces could account for them? Though a set of several different forces might be at work, a single force probably looms large, given how well it correlates with at least two of the three overall patterns we observe: the imbalance between crop agriculture's uptake of nutrients and the application of fertilizers.

The temporal pattern in soil nutrient trends seems to mirror movements in agriculture's demands on tilled soils. The fact that the strongest signs of possible decline in soil organic matter and total nitrogen come

Table 4.8
Chemical fertilizer application per sown hectare, China, 1983

	Kilograms per hectare
Northern China	
Heilongjiang	46.7
Jilin	100.3
Liaoning	180.5
Beijing shi	169.2
Tianjin shi	65.8
Hebei	117.4
Nei Mongol	31.1
Ningxia	60.3
Gansu	51.4
Shaanxi	75.4
Shanxi	75.3
Henan	115.3
Shandong	172.4
Central and southern China	
Jiangsu	174.3
Shanghai shi	210.8
Zhejiang	151.2
Anhui	115.3
Hubei	107.7
Sichuan	111.5
Guizhou	89.6
Hunan	135.0
Jiangxi	85.8
Fujian	196.0
Guangdong	166.2
Guangxi	112.4
Yunnan	97.3
Far west	
Qinghai	77.7
Xinjiang	58.9
Tibet	61.1

Source: Stone 1986, pp. 460–461.

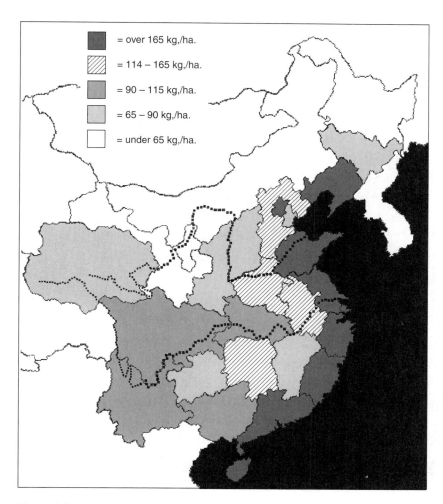

Figure 4.9
Chemical fertilizer application per sown hectare, China, 1983
Source: Stone 1986, pp. 460–461.

after the 1950s suggests a link to the change in agricultural intensity during that time. To judge the intensity of agricultural removal of nutrients, one could look either at yields per acre or at the multiple-cropping index (MCI), or harvests per year, derived by dividing sown area by the area cultivated at any time. Let us prefer the MCI measure here, since the yield measure involves some confusing feedback from the quality of the soil itself. From the 1930s through the late 1950s, the multiple-cropping index (sown area/cultivated area) fell from 1.49 to about 1.39 for China as a whole, with similar movements for the north (Buck 1937, p. 216; Social Science Research Council 1969; Stone 1988). Yields per sown area also grew relatively slowly. Then from the late 1950s the multiple-cropping index rebounded to 1.47 by 1982 and to 1.55 by 1990, and yields also accelerated.

The spatial pattern of the progress of soil nitrogen also seems to be related to the spatial pattern in the intensity of agriculture's demand for nutrients. Table 4.9's display of the MCIs for different regions and provinces suggests the geography of this intensity. In all periods, intensity was greater to the east. In fact, the east experienced greater cropping demands *and* greater application of fertilizer *and* a greater decline in soil nitrogen.

As a working guess, it would appear that the greater crop uptake of nitrogen outweighed the greater application of fertilizer. In effect, the land got a partial rest before the late 1950s but was subjected to rapidly rising gross decomposition of organic matter and nitrogen uptakes thereafter. These uptakes rose at least in step with fertilizer applications. The peculiarity of the southwest fits this suggestion: Its cropping index and its fertilizer application rate were always low by southern standards, and its cropping index clearly rose less than that of the rest of the south after the 1950s. Perhaps the southwest continued to accumulate soil nitrogen content after the 1950s because less was demanded of its soils than of soils further east.

Summary

Some limited historical samples of soil profiles suffice to test some very broad hypotheses about soil trends in China. This chapter has mapped

Table 4.9
Cropping intensity in selected regions and provinces of China, 1933–1983

Agricultural region or province	Multiple cropping index (100 * sown area / cultivated area)		
	1933	1957	1983
Huang-Huai-Hai plain	139	116	146
Hebei		85	131
Henan		131	160
Shandong		133	146
Winter wheat–millet area	118	93	117
Shaanxi		92	127
Shanxi		94	107
Sichuan basin	167	150	181
Sichuan		150	181
South Chang Jiang plain	169	141	234
Hunan		148	228
Jiangxi		143	229
Zhejiang		124	252
Double-cropping area	176	171	193
Guangdong		178	200
Guangxi		148	177
Southwest rice area	152	137	152
Yunnan		136	140
Guizhou		130	153

Sources: Buck 1937, p. 216; Social Science Research Council 1969; Stone 1988; Crook 1992.

Notes: The agricultural regions shown here imitate some of those shown in figure 4.3. For 1957 and 1983 the regions are simply averages of the provinces listed under them, using cultivated areas as weights. Half of Guangxi was allocated to the double-cropping rice region, half to the southwest rice region.

these trends in figures 4.4 and 4.7 and cross-checked them against other evidence. The next chapter will compare the various positive and negative effects these soil trends have had on yields.

China's struggle against excessive soil alkalinity (high pH) and acidity (low pH) has brought mixed results. A victory over alkalinity and salinity shows up clearly in two coastal areas and perhaps in parts of the Huang-Huai-Hai plain and the Wei valley. By contrast, alkalinity and salinity remained severe, without trend, in the Great Bend of the Huang He (Yellow) River and in the lowlands of Jilin and Heilongjiang. In some parts of the desert far west alkalinity and salinity may have worsened, but these areas were already marginal early in this century. In parts of the south, soil acidity abated between the 1930s and 1950s, but part of this gain was reversed between the 1950s and the 1980s.

Trends in China's soil organic matter and macronutrients since the 1930s have shown three broad patterns:

1. *The temporal pattern.* The trends in organic matter, nitrogen content, and acidity were more favorable from the 1930s to the 1950s than later, from the 1950s to the 1980s.

2. *The east-west pattern in nitrogen and organic matter trends.* Between the 1950s and the 1980s, at least, there was a contrast between the stagnation or decline of total soil nitrogen and organic matter toward the east and more favorable trends in the northwest and (especially) the southwest.

3. *The chemical pattern.* The trends were much more positive for total phosphorus and total potassium than for nitrogen and organic matter.

These patterns recommend a change in the search for soil degradation trends in China. Attention should probably shift from erosion to agricultural net depletion of nitrogen and organic matter on erosion-free lands, according to the temporal, spatial, and chemical patterns summarized above.

Agricultural intensity involves a tug-of-war between supplying extra nutrients, from fertilizer and from any shift to nitrogen-fixing crops, and removing more nutrients in plant matter. This intensity correlates with depletion of soil nitrogen but accretion of total phosphorus and potassium. It therefore appears that the net consequence of fertilizer inputs and crop uptake must be quite different for soil nitrogen than for soil phosphorus and potassium. The combination of intense cultivation and intense fertilizer application has left a stagnation or net depletion of total nitrogen and possibly of organic matter but has led to a net buildup of total P and total K. Although one suspects that this buildup takes forms that are mostly unavailable for plant growth in the short run, the buildup of phosphorus and potassium nonetheless may offer an important slow cumulative stimulus to yields. Chapters 5 and 6 will confirm that this appears to be the case for China, and chapter 8 will pick up a similar result for Indonesia.

5

China's Soil-Agriculture Interactions

If China's soil quality deteriorated in some respects and improved in others between the 1930s and the 1980s, what is the net result? Did it deteriorate or improve overall? To ask such a question is to ask for measurements that can transform each soil characteristic into a contribution to overall soil quality. This chapter supplies such measurements for China, using agricultural productivity as the dimension on which we measure soil quality. We will weigh the productivity consequences of each soil characteristic using a cross-section of China's counties (*xian*) in the 1980s and will quantify some feedbacks of agricultural activity on the soil. Chapter 6 will then use the productivity estimates and the history of China's cultivated areas to sketch the trends in China's agricultural soil endowment since the 1930s.

To weigh the conflicting trends in soil quality in economic terms requires some means of putting economic values on each soil characteristic. Since soil quality is an input into agricultural production, it should suffice to let the market price of soil quality reveal its marginal product. In a market setting, the interested scholar might estimate the market price of each soil attribute by hedonic regressions explaining variations in land rents or land values.[1] China lacks a land value series, however. The only way, then, to reveal soil quality's marginal product is by estimating an agricultural production function. This chapter does just that, using cross-sectional data from China's counties in the 1980s within a larger simultaneous-equation system.

Simultaneous Feedbacks: The Conceptual Task

The most formidable barrier to estimating what we have done to the soil, and what difference its condition makes to our agriculture, is the mutuality of the soil-agriculture interactions. People shape the soil that feeds them.[2] This well-known mutuality poses a tough challenge for statistical analysis. It is harder to reach unbiased estimates of two-way causal links between a pair of variables than it is to estimate a one-way causal link. The barriers to estimating how soil affects agriculture and vice versa, and the choices available to us, can be illustrated with a much simplified algebraic statement of the soil-agriculture system involving the following pair of simultaneous equation sets:

Group A: $A = a_0 + a_1 Xa + a_2 S + e_a$,

Group S: $S = b_0 + b_1 Xs + b_2 A + b_3 L(A) + e_s$,

In the first group of equations, Group A, A represents a vector of agricultural outcomes (outputs, crop planting areas). The independent variable Xa is a vector of nonsoil influences on those outcomes, such as human-controlled inputs (labor, fertilizer, irrigation, seed varieties, equipment, draft animals, etc.) and climate. S is a vector of variable soil characteristics that directly affect plant growth, such as soil nutrient levels, organic matter, pH, and topsoil depth; ea is a random error term, and the as are sets of coefficients. This group of equations by itself, ignoring the second group, represents the time-honored production function approach of agricultural economics.[3]

In the second group of equations, Group S, Xs represents a vector of relatively fixed determinants of topsoil quality at the soil site, such as parent material, climate, soil classification, terrain, texture, and topsoil depth. The agricultural outcomes vector reappears twice here, first in its current values (A) and then as $L(A)$, a vector of its lagged values from the past. The lags are indispensable here, because agricultural practice affects soil characteristics only slowly over the years. The bs are coefficients, and es is the error term. If we can estimate the coefficients in both equations, we have unlocked the whole system. Not only will we then know how the truly exogenous forces (Xa and Xs) affect both soil characteristics and agricultural performance, but we can judge the direct effects of soil conditions on performance (a_2) and the direct feedback from agriculture to the soil itself (b_2 and b_3).

The key requirements for empirical success here are measures of all variables that are both reliable and highly variable across some sample. No independent variable can reveal its true influence if it does not vary over the sample, or if it is so closely correlated with another independent variable as to defy sorting out their separate influences.

The ideal experiment for identifying how this system works is easy to describe. Exploring the interactions of, say, ten dimensions of soil quality and ten dimensions of agricultural performance is a manageable task with a sample of more than fifty years' annual data on more than forty land areas if government policies, market conditions, property rights institutions, geology, and climate all vary among these places and years in measurable ways. The information available for China fails to manifest one part of this ideal but does manifest another. It cannot offer that fifty years of annual data. Rather we must deal with a cross section at a single point in time, though as the twenty-first century progresses, it should be increasingly feasible to run tests on pooled time and space samples for China. Nonetheless, within the limitation of a single cross section, China does allow a good look at the separate influences of soil conditions on yields and of exogenous human inputs and policies on soil quality.

Designing a Simultaneous System to Fit China in the 1980s

Agronomy and institutions together define the roles of endogenous and exogenous variables in China's soil-agriculture system. Soil conditions affect more than just the current level of yields: They affect agricultural inputs as well. In the longest-run perspective, one could view all nonsoil inputs as endogenous by-products of soil geology and climate that over the centuries govern a region's whole population level and agricultural input intensity as well as its outputs. Thus one could, in figure 5.1's schematic sketch of causality in the countryside, see everything as flowing from the soil and climate forces featured on the left.

Yet events since the 1930s have largely shaped China's agricultural geography, particularly the communist victory in 1949, the waves of policy revolution under Chairman Mao Tse-Tung, and the reforms since

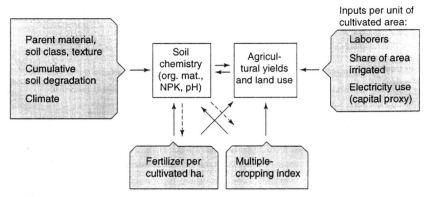

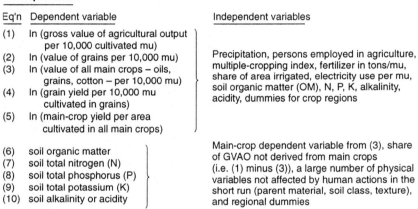

The equations:

Eq'n   Dependent variable   Independent variables

(1)   ln (gross value of agricultural output
         per 10,000 cultivated mu)

(2)   ln (value of grains per 10,000 mu)   Precipitation, persons employed in agriculture,
(3)   ln (value of all main crops – oils,   multiple-cropping index, fertilizer in tons/mu,
         grains, cotton – per 10,000 mu)   share of area irrigated, electricity use per mu,
(4)   ln (grain yield per 10,000 mu   soil organic matter (OM), N, P, K, alkalinity,
         cultivated in grains)   acidity, dummies for crop regions
(5)   ln (main-crop yield per area
         cultivated in all main crops)

(6)   soil organic matter   Main-crop dependent variable from (3), share
(7)   soil total nitrogen (N)   of GVAO not derived from main crops
(8)   soil total phosphorus (P)   (i.e. (1) minus (3)), a large number of physical
(9)   soil total potassium (K)   variables not affected by human actions in the
(10)   soil alkalinity or acidity   short run (parent material, soil class, texture),
                                                     and regional dummies

Figure 5.1
Causal influences in a cross section of China's soils and agricultural products,
1980s

1978. It is only a slight exaggeration to say that China between 1949 and
the 1990s lacked not only a land market but also a labor market, a capital
market, and a fertilizer market. The tight restrictions on labor mobility
have more or less frozen China's demographic geography in place. Until
the mid-1980s, the same restrictions have even kept farm labor from
shifting to nonagricultural production. In a spatial crosssection, the number of laborers available for agriculture is therefore an exogenous variable, as shown by its position on the right-hand edge of figure 5.1. Other
inputs are relatively exogenous. Irrigation is driven by terrain, by the

history of public construction campaigns, and by the centralized allocation of tangible nonland capital, here represented by electricity use per unit of cultivated land. The spatial allocation of chemical fertilizers was dictated by central government and provincial policy before the 1990s. Fertilizer was a resource for which provincial governments vied at the national level and county and local governments vied at the provincial level. A more detailed version of the present model uses province and county attributes as determinants of the lobbying power of subgovernments in augmenting their supplies of subsidized fertilizer. Similarly, the intensity with which laborers and other inputs are thrown into multiple cropping also depends on the policy environment.[4]

At the center of this causal system, soil conditions and agricultural outputs and inputs are determined jointly. The agricultural production function is assumed to take the following general form:

$$\ln(Y/L) = a_0 + W \text{ (precipitation, temperature)}$$
$$+ T \text{ (}Lab, MCI, Fert, Irrig, Elec\text{)}$$
$$+ Q \text{ (}OM, N, P, K, Alk, Acid\text{)}, \tag{1)–(5}$$

where Y = either crop output or the gross value of agricultural output, L = cultivated land area, a_0 is a productivity shift parameter, W is a yield-relevant weather or climate index, T is a translog production function, and Q is a soil quality function. The yieldlike dependent variable $\ln(Y/L)$ comes in five variants:

$\ln(Y_{ag}/L_{tot})$ = log of the gross value of agricultural output (GVAO) per 10,000 mu of total agricultural cultivation (15 mu = 1 ha.); $\qquad$ (1)

$\ln(Y_{grain}/L_{tot})$ = log of the value of grains per 10,000 mu of total cultivated area; $\qquad$ (2)

$\ln(Y_{main}/L_{tot})$ = log of the value of main crops (grains, oils, and cotton) per 10,000 mu of total cultivated area; $\qquad$ (3)

$\ln(Y_{grain}/L_{grain})$ = log of the value of grains per 10,000 mu of area cultivated in grains; $\qquad$ (4)

$\ln(Y_{main}/L_{main})$ = log of the value of main crops (grains, oils, and cotton) per 10,000 mu of area cultivated in the main crops. $\qquad$ (5)

Note that the "cultivated land area" denominator in the first three equations is total cultivated area for all of agriculture, not a land area specific to grains or to other main crops. For the crops in equations (2) and (3), each yield variable is a yield per hectare cultivated in those crops times their share of total cultivated land area. That is, the dependent variable in either (2) or (3) is a yield times a land share.[5]

In the translog function (T), *Lab* is just a body count of persons employed in agriculture in 1985, per cultivated hectare. To get still closer to true labor inputs, we add the multiple cropping index, ln (MCI) = sown area/cultivated area (L), to the list of variables, as if it were a separate dimension of total labor input. We also add the (logs of) tonnage of chemical fertilizer (*Fert*), the irrigated share of total cultivated area (*Irrig*), and electricity use per unit of cultivated area (*Elec*).

The production function T is translog such that $\ln(T)$ is a function of the log of each input, of its square, and of interaction terms involving pairs of logged inputs. We shall alter the set of interactions slightly. Since initial regressions showed no significant interaction between the multiple-cropping index (*MCI*) and the other inputs, we drop these interaction terms. On the other hand, in equations (2) and (3), where the dependent variable of the overall equation is yield only from specific crops and not total agricultural output per hectare, we add terms that account for the interaction of each input with the share of crop output in total agricultural product:

$$T = c_1 \ln Lab + c_2 \ln MCI + c_3 \ln Fert + c_4 \ln Irrig + c_5 \ln Elec + c_6 \ln^2 Lab + c_7 \ln^2 MCI + c_8 \ln^2 Fert + c_9 \ln^2 Irrig + c_{10} \ln^2 Elec + c_{11} (\ln Lab \cdot \ln Fert) + c_{12} (\ln Lab \cdot \ln Irrig) + c_{13} (\ln Lab \cdot \ln Elec) + c_{14} (\ln Fert \cdot \ln Irrig) + c_{15} (\ln Fert \cdot \ln Elec) + c_{16} (\ln Irrig \cdot \ln Elec). \qquad (1a)$$

If the true production function is Cobb-Douglas or constant elasticities of substitution (CES), the translog specification will reveal this, since the Cobb-Douglas and CES are special cases nested in the translog function used here.

Exogenous policy and history combined to dictate the supplies of total labor, electricity use, and irrigated land area available to each county, the unit of observation used here. Policy and history even governed the allocation of labor and land area among grain, other main crops, and all other agricultural products. Two of the agricultural input variables—chemical-fertilizer application (ln *Fert*) and the multiple-cropping index (ln *MCI*)—might conceivably be endogenous, as mentioned. They might respond to soil chemistry, which affects both the incentive (need) for chemical fertilizers and the returns from multiple cropping. This possibility, represented by the dashed arrows in figure 5.1, was tested in equations reported elsewhere. Although those equations showed that fertilizer and multiple cropping are significantly related to policy variables, the feedbacks from soil chemistry to fertilizer use and multiple cropping were insignificant, allowing us to view fertilizer and multiple cropping as exogenous to the system shown in figure 5.1.[6]

In the soil quality function (Q), *OM, N, P,* and *K* are the natural logs for organic matter, total nitrogen, total phosphorus, and total potassium, respectively. The natural log specification reflects the belief that the marginal product of absolute nutrient levels declines, but remains positive, as those nutrient levels rise. In addition, some interaction terms will be added to explore some effects predicted by agronomists and soil scientists, namely, interactions between the yield effects of organic matter and phosphorus, between organic matter and clay content, between phosphorus and pH, between phosphorus and clay content, and between potassium and acidity.[7] Agronomic research suggests that the direct effects of *Acid* and *Alk* are also nonlinear, in the sense of being bounded by zero: *Acid* = max $(0, 6 - \text{pH})$ and *Alk* = max $(0, \text{pH} - 8)$. *Alk* is an important endogenous variable for north China, as is *Acid* for the south.

On the other side of the soil-agriculture interaction, each of the soil parameters is a predicted value, because it in turn depends on agricultural practice:

$$OM = OM \ (Fert, \ Irrig, \ Y_{main}/L_{tot}, \ (Y/L)_{other}, \ MCI, \ Z), \tag{6}$$

$$N = N \ (Fert, \ Irrig, \ Y_{main}/L_{tot}, \ (Y/L)_{other}, \ MCI, \ Z), \tag{7}$$

$$P = P \ (Fert, \ Irrig, \ Y_{main}/L_{tot}, \ (Y/L)_{other}, \ MCI, \ Z), \tag{8}$$

$$K = K \ (Fert, \ Irrig, \ Y_{main}/L_{tot}, \ (Y/L)_{other}, \ MCI, \ Z), \tag{9}$$

$$Alk = Alk \ (Fert, \ Irrig, \ Y_{main}/L_{tot}, \ (Y/L)_{other}, \ MCI, \ Z) \text{ in the north, or}$$

$$Acid = Acid \ (Fert, \ Irrig, \ Y_{main}/L_{tot}, \ (Y/L)_{other}, \ MCI, \ Z) \text{ in the south,} \tag{10}$$

where the yield of products other than the main crops is represented by the log-difference

$$\ln \ (Y/L)_{\text{other}} = \ln \ (Y_{ag}/L_{tot}) - \ln \ (Y_{main}/L_{tot}),$$

which allows us to examine the effects of shifting China's product mix toward or away from the main staple crops, effects that turn out to be central to a discussion of likely future trends; and Z is a vector of other determinants of soil quality discussed in chapter 4, including soil class, parent material, texture, land use (in some cases), and median depth of the sampled horizon.

We have, then, a simultaneous-equation system in which soil and agriculture interact, consisting of equations (1)–(10).[8]

The best available data set for exploring these interactions matches soil profile data for 1981–1986 with agricultural outputs and inputs in 1985 for a sample of counties (*xian*).[9] The places in the two data sets do not match perfectly, in that the profiles are taken in particular locations within each county. The approach taken here is to use the sample on which the dependent variable is measured. Thus equations (1)–(5), explaining counties' agricultural behavior, are run on county data grouped by 161 northern and 282 southern counties, with soil characteristics averaged over the soil profiles within each county. Equations (6)–(10), by contrast, are run on soil profile samples, with 266 northern and 495 southern soil profiles. One must take care to interpret these results as cross-sectional, not as an experimental or historical time series. That is, the effect of x on y here means "how places where x occurred differed from other places in their y, for both human and other reasons," not "how changing x on the same plots changed y from one period to the next."

The Determinants of Agricultural Yields

Tables 5.1 and 5.2 report some elasticities from two-stage regressions on the simultaneous system sketched here and in figure 5.1. Appendix C gives a fuller reporting of the regression results. Let us survey the elasticities in the order of the sets of equations, starting in table 5.1 with equations (1)–(5).[10] At the top of equation (1), we find that raising any of the five nonland agricultural inputs raises the gross value of agricultural output (GVAO) in both parts of China, as expected. The strongest positive effect comes from labor inputs, in the form of either extra persons or more multiple cropping. Comparing across the five equations in table 5.1 shows that the positive input effects include effects on crop-specific yields as well as effects on the allocation of cultivated area among product categories.

The responses of yields to soil chemistry, the component parts of the soil quality function Q, are less consistent by region and product than the effects of the nonland inputs. The clearest soil effect in table 5.1 is that of extreme soil pH. Both alkalinity in the north and acidity in the south clearly reduce total GVAO. They do so without significantly reducing crop-specific yields of grain, oils, and cotton, however (in (4) and (5)). Rather the effect is on secondary-product yields, combined with a higher share of cultivated (and sown) areas taken by those secondary products.

Other soil chemistry parameters have weaker, or more regional, effects on yields. Organic matter and nitrogen need to be viewed together, because they tend to cycle around an equilibrium OM/N ratio generally estimated at between seventeen and twenty to one. Together they have no significant *marginal* effect on overall yields, either north or south. Presumably, the lack of significance results from humans' ability to substitute quick-release nitrogen for soil nitrogen, with similar implicit substitution in organic matter.[11]

Extra soil potassium was a slightly positive influence on crop production in the south, where average potassium levels are lower. Total soil phosphorus tends to have a puzzlingly significant negative effect, though a small one, on agricultural productivity in the south. One would have predicted a more positive role for phosphorus in the south. The negativity does not mean, of course, that greater phosphorus reduces plant growth. Rather, it seems to mean that in some southern soils where total phosphorus is higher, for geological reasons eluding our measures, the phosphorus is tied up in relatively less plant-available forms such as apatite minerals.

An important pattern emerging here is that China's agricultural yields seem to depend mainly on those soil chemistry parameters that in fact have not deteriorated since the 1930s, especially alkalinity in the north and acidity and potassium in the south. Agricultural output in south China clearly depends on acidity, which was reduced between the 1930s and the 1950s (but may have worsened somewhat in parts of the south after the 1950s). Alkalinity assumes a similar, but less prominent, role in

Table 5.1
Influences on agricultural yields in China, 1985

Independent variables	Equation numbers and dependent variables									
	(1) ln (GVAO per cultivated area)		(2) ln (grain per total cultivated area)		(3) ln (main crops per total cultivated area)		(4) ln (grain yield per grain-cultivated area)		(5) ln (main-crop yield per area cultivated in main crops)	
	Coeff.	$|t|$	Coeff.	$|t|$	Coeff.	$|t|$	Coeff.	$|t|$	Coeff.	$|t|$
North China										
ln (laborers/cultivated area)	0.350	(4.00)	0.317	(4.15)	0.199	(2.20)	1.116	(6.55)	2.066	(3.85)
ln(multiple-cropping index)	0.516	(1.99)	0.045	(0.20)	0.105	(0.39)	-0.525	(1.04)	0.197	(1.13)
ln (fertilizer/cultivated area)	0.280	(4.30)	0.365	(6.56)	0.315	(4.52)	0.352	(2.77)	0.493	(0.27)
ln (irrigated share of cultivated area)	0.096	(1.89)	0.133	(2.84)	0.181	(3.41)	-0.015	(0.15)	0.286	(1.08)
ln (electricity/cultivated area)	0.069	(1.82)	0.033	(1.02)	0.011	(0.27)	0.082	(1.11)	0.286	(1.22)
ln (organic matter)	-0.052	(0.38)	-0.175	(1.48)	-0.139	(0.99)	0.034	(0.13)	0.070	(0.26)
ln (total nitrogen)	-0.0003	(0.00)	0.099	(0.70)	-0.058	(0.35)	-0.065	(0.21)	-0.252	(0.78)
ln (total phosphorus)	-0.031	(0.63)	-0.017	(0.40)	-0.051	(1.01)	0.145	(1.50)	0.110	(1.11)
ln (total potassium)	-0.023	(0.19)	0.091	(0.90)	0.215	(1.74)	-0.443	(1.91)	-0.263	(1.11)
Alkalinity	-0.267	(1.59)	-0.150	(1.04)	-0.383	(2.16)	0.009	(0.03)	-0.146	(0.44)
South China										
ln (laborers/cultivated area)	0.298	(7.17)	0.124	(3.33)	0.119	(3.23)	0.512	(4.02)	0.502	(4.01)
ln(multiple-cropping index)	0.127	(1.35)	0.584	(6.65)	0.577	(6.87)	0.235	(0.82)	0.244	(0.87)

Table 5.1
(continued)

Independent variables	Equation numbers and dependent variables									
	(1) ln (GVAO per cultivated area)		(2) ln (grain per total cultivated area)		(3) ln (main crops per total cultivated area)		(4) ln (grain yield per grain-cultivated area)		(5) ln (main-crop yield per area cultivated in main crops)	
	Coeff.	|t|	Coeff.	|t|	Coeff.	|t|	Coeff.	|t|	Coeff.	|t|
ln (fertilizer/cultivated area)	0.157	5.39	0.125	(4.63)	0.139	(5.34)	0.054	(0.61)	0.059	(0.67)
ln (irrigated share of cultivated area)	0.165	(5.02)	0.242	(7.94)	0.271	(9.45)	0.055	(0.55)	0.067	(0.68)
ln (electricity/cultivated area)	0.028	(1.89)	−0.007	(0.49)	−0.016	(1.21)	0.131	(2.84)	0.126	(2.78)
ln (organic matter)	0.823	(4.88)	0.202	(1.32)	0.211	(1.45)	0.134	(0.26)	0.186	(0.37)
ln (total nitrogen)	−0.753	(4.37)	−0.227	(1.44)	−0.321	(2.14)	0.130	(0.25)	0.030	(0.06)
ln (total phosphorus)	−0.060	(2.58)	−0.069	(3.36)	−0.054	(2.68)	−0.192	(2.70)	−0.178	(2.55)
ln (total potassium)	0.090	(2.03)	0.179	(4.45)	0.147	(3.81)	0.136	(1.00)	0.121	(0.91)
Acidity	−0.227	(3.57)	−0.046	(0.79)	−0.190	(3.31)	0.272	(1.40)	0.171	(0.89)

Notes: All outputs are priced in grain equivalents, as explained in the notes to appendix tables C.1 and C.2.

GVAO—gross value of agricultural output, as in the text.

Main crops—grains plus oils plus cotton, valued in grain equivalents.

Each coefficient is an implied slope at the mean values of the independent variables.

Equations are second-stage estimating equations corresponding to equations (1)–(5) and (6)–(10) in the text and figure 5.1.

For fuller regression results and variable definitions, see appendix table A.5.

For complete regression printouts of the variant where fertilizer and multiple cropping are instrumented, see Lindert 1996.

Table 5.2
Influences on soil chemistry in cultivated soils of China, 1980s

Independent variables	Equation numbers and dependent variables							
	(6) ln (percentage organic matter)		(7) ln (total nitrogen)		(9) ln (total potassium)		(10) Alkalinity = max (pH − 8, 0)	
	Coeff.	|t|	Coeff.	|t|	Coeff.	|t|	Coeff.	|t|
North China								
Agricultural outputs								
ln (main crops/cultivated area)	0.199	−(1.29)	0.002	(0.01)			−0.489	(2.44)
ln(all output/main crops)	0.636	(3.58)	0.423	(2.44)			−0.111	(0.52)
Agricultural input intensities								
ln (multiple-cropping index)	−0.387	(0.42)	−0.347	(0.39)			−1.130	(1.23)
ln (fertilizer/cultivated area)	−0.118	(1.43)	−0.006	(0.07)			0.009	(0.12)
South China								
Agricultural outputs								
ln (main crops/cultivated area)	−0.524	(2.48)	−0.774	(3.87)	0.042	(0.16)	−0.108	(0.59)
ln(all output/main crops)	0.344	(1.73)	0.180	(0.96)	0.024	(0.10)	−0.591	(1.66)
Agricultural input intensities								
ln (multiple-cropping index)	−0.150	(0.77)	0.054	(0.29)	−0.249	(1.03)	0.213	(0.59)
ln (fertilizer/cultivated area)	−0.052	(0.80)	−0.071	(1.16)	0.146	(1.83)	−0.036	(0.31)

Notes: See the notes to table 5.1.
Equations omitted are those that failed the *F*-test, leaving the dependent variable exogenous for present purposes. For north China, equations (8)[a]nd (9) for phosphorus and potassium are treated as exogenous and omitted. For south China only equation (8) for phosphorus is exogenous and omitted.

relating trends to productivity impacts: Alkalinity showed no clear trend in the north, yet its level does matter to productivity.

By contrast, as we have seen, organic matter and nitrogen do not seem to matter much at the intercounty margin. How could this be, given the known dependence of plant growth on available nitrogen and on the various benefits of higher organic matter content, and given that China's topsoils are generally low in these crucial soil endowments? What makes this possible is that quick-release nitrate fertilizers, such as urea, can act as a substitute for soil organic matter and total nitrogen at the margin, even though minimum levels must be maintained. Importantly, then, the topsoil nitrogen depletion since the 1950s in the Huang-Huai-Hai plain and the Chang Jiang (Yangzi) plain may have been offset relatively easily. We return to this issue below.

Feedbacks from Agriculture to the Soil

On the other side of the causal circle linking soil to agriculture, table 5.2 explores the effects of crop yields, fertilizer and cropping intensity on the soil itself. Each elasticity-at-means is based on a log-quadratic form from equations (6)–(10) to allow for possible curvature in its effect on the soil.

Of the five equations, those for phosphorus are insignificant in both regions, as are those for potassium in the north. By implication, phosphorus and potassium are exogenous in northern China, as in phosphorus in the south. The remaining soil equations reveal important patterns.

Raising the yields of China's main crops (grains, oils, and cotton) does the topsoil endowment few favors, according to cross-sectional patterns in the 1980s. True, in the north, the areas with greater main-crop outputs also achieved lower soil alkalinity. This benefit is probably not just a mirage created by reverse causation, since the negative effect of alkalinity on yields and crop choice was already addressed in the two-stage estimation of equations (1)–(3). Yet cutting northern alkalinity is the only clear soil benefit from human interventions that produce more of the main crops. Growing more grains, oils, and cotton did not seem to raise soil organic matter or nitrogen in the north in the 1980s. Nor did it significantly alleviate soil acidity in southern China. Worse, growing more of the main crops reduced soil organic matter and nitrogen in the south.

By contrast, raising the yields of agricultural products other than the main crops had clearly beneficial soil effects, as shown in the rows for "ln (all output/main crops)." Extra non-main-crop products, such as vegetables and animal products, raise soil organic matter throughout China, and also tend to raise nitrogen levels and reduce alkalinity and acidity.[12] Any force that shifts land area and other inputs away from grains, oils, and cotton to less staple crops therefore seems to improve the soil characteristics measured here. One aspect of human intervention, namely the output mix, may have strong effects on soil chemistry. Increasing cultivation of grains, oils, and cotton may fail to improve soil endowments (except for alkalinity reduction in the North), whereas shifting to other crops and to animal products may have beneficial effects on soil endowments.

The contrast between increasing production of grains, oils, and cotton and producing other agricultural products implies a possible Engel effect that has gone unnoticed: Could it be that general economic development improves the soil by pulling food demand and input supplies away from producing staples, especially staple grains? It could do so even without anybody's having imagined such a pattern. This is not a certainty, however. Among the tasks needed further research is sorting out the complex implications of rising amounts of animal products for the pattern of land use. Does it really shift land use toward soil-conserving grasslands and legumes, or does it simply raise the nutrient uptake from the soil through conventional nonsoy grains as fodder? We return to this possibility in chapter 9's summary of the possible feedbacks from agriculture to the soil.

The next immediate task, however, is to use the estimates of the productivity impacts of soil characteristics from tables 5.1 and 5.2 to sketch what has happened to China's productivity-relevant soil quality since the 1930s. chapter 6 takes on this task.

6

The Quality and Quantity of China's Cultivated Soils

Although the estimates presented in chapters 4 and 5 are preliminary, they do help answer the question of whether China's agriculture has depleted its soil endowment since the 1930s. To formulate a tentative answer to this question, this chapter proceeds in four steps. First it sketches the trends in soil quality that are suggested by combining chapter 4's trends in soil characteristics with chapter 5's estimates of the marginal impact of each characteristic. Then it weighs the controversial evidence of trends in China's cultivated area and develops a usable set of estimates for cultivated area at the province level. Next it combines the soil quality trends and the cultivated-area trends into a preliminary view of China's overall net investment in its soil endowment since the 1930s. Finally it addresses three concerns that these national trend estimates may seem to have hidden: urban encroachment on China's farmlands, desertification, and land uses that may have increased the dangers of flooding.

Soil Quality since the 1930s: A First Guess

Once we have put a price or value on each soil characteristic (nitrogen, phosphorus, potassium, organic matter, acidity, and alkalinity), we can gain a rough idea of China's net soil quality investment. We can multiply the unit price of each characteristic by the quantity of that same charac-teristic to yield a value of that attribute, measured in purchasing power. These purchasing-power values can then be added up to measure a total investment in productive soil characteristics, the same way we add up the values of new factory buildings and homes and roads to get the econo-mist's usual measure of national investment.

In a market economy the marginal value of each soil characteristic could be inferred from its effect on either farmland rents or farmland purchase prices, as noted in chapter 5. If feasible, this land value approach is a superior technique for valuing soil characteristics. It gets around our scholarly inability to obtain data on all the soil characteristics relevant to productivity by using farmers' own more detailed knowledge about each local soil as revealed by the price they were willing to pay for that particular plot.

For China, as noted in chapter 5, there were no systematic rental or purchase prices of agricultural land before the late 1990s. Lacking the price metric, we must use an alternative method for valuing effects on productivity. Chapter 5 resorted to a production function measure of the marginal product of each soil characteristic. Combining each characteristic's marginal product with its change over time answers the question "What was it worth for China's agriculture in the 1980s to use its 1980s soil endowment instead of its endowment from the 1930s or 1950s?"

Beneath the simple idea of multiplying each soil characteristic's marginal product by its total abundance in China lies a labyrinth of detailed calculations and supporting assumptions. It will suffice here to describe the overall spreadsheet journey in very general terms and to add a caveat before hazarding some guesses at the end of that journey.[1]

The steps toward valuing China's overall soil quality and her land endowment start with soil quality alone and with the 1980s alone. The underlying calculations use the province as the most manageable unit of aggregation across soil class land areas and use Jiangsu = 1.000 as the comparison base for all other provinces. Multiplying chapter 4's estimates of average soil characteristics in the 1980s by chapter 5's marginal productivities of soil characteristics yields a relative soil quality measure for each province in the 1980s (still excluding the far west and the islands).

To value the changes in China's soil quality between the 1930s and 1950s and the 1950s and 1980s calls first for a distribution of each province's soils among the agricultural regions introduced in figure 4.3, which makes it possible to construct province averages for the changes in soil characteristics across the decades. Since the 1980s the soil data are much richer than those of the earlier decades and everything is compared to the 1980s, more specifically, to the average soil quality for Jiangsu province in the 1980s. Thus the net change in organic matter is weighed for some provinces as a 1980s to 1930s comparison, and for an overlapping set of provinces as a 1980s to 1950s comparison. Any comparison of the 1930s with the 1950s is only indirect and is based on thinner data. Once the time comparisons are made and the changes in soil characteristics valued, these changes are aggregated into overall net changes in soil quality for north China, for south China, and for the nation as a whole. Interactions between these final changes in soil quality and the changes in cultivated area then yield estimates of a half century of changes in China's soil quality and net soil endowment.

The main caveat about the exercise is that the confidence interval around each of its estimates will be unknown and, in some sense, wide. The interval is unknown because each estimate is a sum of products, a functional form that defies simple aggregation of standard errors. The products themselves involve both regression coefficients, for which we have separate standard errors from chapters 4 and 5, and some land areas for which we have no standard errors of measurement. The confidence interval is probably wide in that it builds on estimates that can be wildly off, as were the official figures on cultivated area before (and since?) the finding that China's cultivated area is roughly 38 percent higher than previously reported. One can only say that the results seem to be a fair first guess at the net changes in China's soil quality and land endowment.

With this caveat, we can view the apparent historical changes in China's soil quality in panel A of table 6.1. The average productivity of China's topsoils has probably not declined since the 1950s and perhaps not even since the 1930s. For the 1930s the paucity of data limits the measurement of soil quality covering all of our main soil characteristics. Table 6.1 gives a comprehensive measure only for two agricultural regions (the winter wheat–millet region and the north Chang Jiang plain), even though chapter 4 offered more 1930s averages for individual characteristics.

From the 1950s to the 1980s, we have enough detail from several regions to place some credence in table 6.1's finding that average soil quality did not decline. Over these decades, the value of China's topsoil characteristics apparently rose 1.0 percent in the north, 19.5 percent in

Table 6.1
Rough measures of China's net changes in soil quality and cultivated area, 1930s–1990s

Panel A. Soil quality changes (excluding islands and desert mountain west)

Aggregate changes in soil quality from the 1930s to the 1980s (Jiangsu 1985 = 100):

	1930s level	1980s level	Change (%)
North China, 1 region	95.6	99.7	4.3
South China, 1 region	97.9	95.0	-3.0
All China, 2 regions	96.6	96.2	-0.4

Aggregate changes in soil quality from the 1950s to the 1980s (Jiangsu 1985 = 100):

	1950s level	1980s level	Change (%)
North China, 4 regions	97.5	98.5	1.0
South China, 4 regions	79.5	95.0	19.5
All China, 8 regions	90.4	97.2	7.4

Panel B. Changes in agricultural land areas (all China, millions of hectares)

	1933	1957	1978	1985	1993	1995
Usual cultivated area series	98.9	111.8	99.3	96.7	95.0	95.0
Alternative cultivated area series	109.0	111.8	137.0	133.5	131.1	131.1
Heilig's cultivated area estimates				139.7		137.1
All cropland + pasture (FAO)		323.0	419.4	468.3	496.0	

Table 6.1
(continued)

Panel C. Correlations between unreported-land share (Wang et al. 1992, cited in Heilig 1997) and other province attributes, 1985

With soil quality in 1985 (as in panel A):	−0.163
With rural income per capita in 1988	−0.285
With agricultural production per capita in 1986	−0.638

Notes: Crude estimates presented here are based on detail at the level of the province/agri-region, as detailed in Lindert 1996.

Agricultural regions, as mapped in Buck 1937, p. 27, and in figure 4.3, are as follows:
North China, 1 region—winter wheat–millet region
South China, 1 region—north Chang Jiang plain
North China, 4 regions—spring wheat regions northwest and northeast, Huang-Huai-Hai plain, saline coast
South China, 4 regions—north and south Chang Jiang plain, double-cropping rice region, southwest rice region.

The procedure for estimating total land endowments applies marginal log-products of land from table 5.1 as unit log-values for combining different soil characteristics into a single-value real-value measure. Regions are aggregated up from province detail using 1985 cultivated areas as weights. The same cultivated-area weights are used to make north and south China aggregates and national aggregates.

All cultivated areas for 1933 and 1957 are drawn from Chao 1970, pp. 194, 281, except that Ningxia 1957 is a 1965 figure from Dawson 1970, p. 53. One hectare = 15 mou.

Usual cultivated area series: For 1933 the estimate is "Buck IV, Computation II" as quoted in Chao 1970, p. 193. Official figures for 1978–1993 are based on the official agricultural statistical yearbook, as cited in a data file kindly supplied by Justin Lin.

Alternative cultivated area: For 1933, this equals the Buck IV figure plus 152.74 million mou in cultivated lands that escaped registration before 1957 (see Chao 1970, pp. 193–208). For 1978–1993, the alternative cultivated-area figures are the usual (official) ones times 1.38, based on the satellite evidence that underregistration has hidden 38 percent of China's cultivated lands (Crook 1992). This share of land is assumed to have gone into hiding between 1957 and 1978 and to have remained constant as a share of cultivated land since then.

Heilig's estimates are from Heilig 1997, pp. 139–45, which also discusses the competing estimates for the 1980s and 1990s.

All cropland + pasture—FAO's "arable and permanent crops" plus "permanent pasture" from its annual *Production Yearbook*. This excludes forest and woodlands and all other lands (cities, deserts, etc.)

For 1988 rural income per capita (not just agricultural income): Khan et al. 1988, p. 53.
For 1986 gross value of agricultural product per person in agriculture: Field 1993, pp. 152–53.

Table 6.2
Regional indices of soil quality, 1950s–1980s

	Soil quality index (Jiangsu 1985 = 100)		
Region (see figure 4.3)	1950s	1980s	Percentage change
North Chang Jiang plain	77.9	95.0	+22
South Chang Jiang plain	82.9	93.5	+13
Double-cropping rice	91.0	93.4	+2
Southwestern rice	68.6	98.1	+43

the south, and 7.4 percent overall, contrary to the widespread belief in a net loss in soil quality.[2]

The net change in soil quality from the 1950s to the 1980s thus looks positive for the south and about zero for the north. Yet the underlying estimates are not so sturdy that these particular percentage figures should be relied on. In particular, the gain in the south may be overstated here.

The underlying regional estimates are shown in table 6.2. In two of these four southern regions, the North Chang Jiang plain and the South Chang Jiang plain, the net gain in soil quality rests on an oddity. In these two regions, the organic matter to nitrogen ratio rose from the 1950s to the 1980s. As chapter 5 has noted, it is difficult to explain why the positive productivity effect of organic matter should be offset by an almost equally negative effect of nitrogen. That should matter little as long as the organic matter to nitrogen ratio is constant. Yet in the Chang Jiang plain from the 1950s to the 1980s, nitrogen levels dropped while organic matter levels did not. This combination produces a regression-based gain in soil productivity and soil quality in the Chang Jiang plain that defies interpretation.

Thus for south China as a whole, we have soil quality measures that suggest some net improvement in the double-cropping far south and in the southwestern rice region but leave doubts about those gains of 22 and 13 percent for the two halves of the Chang Jiang plain. The safest tentative conclusion is that soil quality for the south as a whole may have risen hardly at all, or may have risen by the 19.5 percent implied by table 6.1. For the whole of China, this means that soil quality may have risen hardly at all, or may have risen by something like that 7.4 percent in table 6.1. The most reliable, or least unreliable, basic inference is that China's overall soil quality did not decline between the 1950s and the 1980s.

Cultivated Area

If our best first guess is that the average quality of China's agricultural lands did not decline in the period under study, what about the quantity of lands available for agriculture? Has the total cultivated area not shrunk, as warned by the works of Lester Brown, Vaclav Smil, and others cited in chapters 1 and 2? Indeed even the usual—in recent times, the official—time series on China's cultivated area say that the total cultivated area declined significantly from the 1950s to the 1980s. China's overall land endowment might have declined, then, even if the average quality of the soil did not.

Yet the figures on cultivated area are themselves controversial. Fortunately, the controversy is being resolved, and a careful handling of the numbers yields a fairly clear (and startling) set of results about changes since the 1930s or since the 1950s. Let us turn first to the reassessment of China's land stock in the 1980s and 1990s before reinterpreting what happened over the earlier decades.

The Cultivated-Area Debate in the 1980s and 1990s

The case for concern over the amount of cultivated area would seem at first to draw support from the official figures, which show an 11 percent decline in cultivated area from 1957 to 1978 and a further 4.3 percent decline from 1978 to 1995, yielding an overall drop of 15 percent of the lands cultivated in 1957.

Yet we now know that the official figures omit a large proportion of unreported cultivated lands. Drawing on satellite data and other clues, many authorities now believe that the cultivated area should be 38 percent, or even 44 percent, higher than the official figures have shown (Crook 1992; Wang et al. 1992; Wu and Guo 1994; Heilig 1997). Even the official statistical yearbook itself has, for several years, posted a warning that "[f]igures for the cultivated areas are under estimated and must be further verified" (China, State Statistical Bureau 1996, p. 355).

With similar sobriety, China's Office of the National Soil Survey (1996, chapter 4) now quietly reports a total cultivated area for the mid-1980s of 130 million hectares (1,950 million mou), much closer to the alternative estimate of 133.5 million hectares for 1985 (Crook 1992, citing Wang and others) than to the official 1985 figure of 96.7 million hectares (1,451 million mou). A new consensus has emerged: China is cultivating much more land than the official annual estimates or the pessimistic scholars have announced. Panel B of table 6.1 summarizes this revisionism with two rows of revised estimates, one that marks up the official estimates by the 38 percent suggested from satellite data and one that reports correlations by Heilig.

Beyond correcting the cultivated area series, one should also note that the concept of cultivated area itself is narrower than the total land area in use by agriculture. China's measures of cultivated area exclude much of the land base used for animal product agriculture, such as pastures, ponds, and semiforest. Including them, as in the estimates by the Food and Agricultural Organization shown in table 6.1, yields a much greater land area available for agriculture, an area that has reportedly grown by 53 percent since 1957. Although this broader land area may include hectares of lower average productivity than the usually measured cultivated area, we are alerted to the possibility that China's land base is much greater than many have believed.

Two immediate questions arise if the revised view is correct, as it appears to be: (1) Are the unreported lands of the same quality as the reported areas, so that we can infer that the true quality-adjusted stock of agricultural lands should be 38–44 percent higher? (2) How recently did this underreporting occur? Did it creep in since 1978? Or has the underreporting been receding recently, so that the true land stock has fallen at least as fast as the pessimists feared?

What Kinds of Cultivated Lands Were Unreported?
Panel C of table 6.1 offers an indirect clue to the quality of the unreported cultivated areas, making use of the fact that the share of cultivated land that went unreported has varied greatly among the provinces. In Guizhou, Ningxia, and Yunnan, more than half the truly cultivated land was unreported, versus less than a fifth in Shanghai, Hebei, Jiangxi,

Jiangsu, and Hubei. The fifth column in table 6.3 underlines these differences with a "correction factor" and the sixth reports the share of lands that were unreported in each province in 1985, as implied by this correction factor. One geographic pattern that emerges is that the share unreported rises with the sum of the distances of the province's center of population from Beijing and Shanghai. Another geographic pattern, supported by a correlation of −0.638 in table 6.1, is that provinces with poorer agriculture managed to underreport their cultivated areas and by doing so presumably paid lower implicit taxes through the quotas system. The correlation is much weaker with a given county's rural, as opposed to agricultural, income.

The weakest relationship is with the average soil quality for the province, which was calculated by combining average provincial values of soil chemical parameters (organic matter, nitrogen, phosphorus, potassium, alkalinity or acidity) with coefficients for their impact on the gross value of agricultural output. As far as the quality of the soil is concerned, there is only a slight tendency for cultivated land to go unreported more in poor-soil than in good-soil provinces. The kinds of lands more consistently underreported, and implicitly subsidized, have been lands worked by poor people, not lands that are themselves of poor quality.[3]

The geographic pattern of underreporting seems to fit a simple theory of how local officials have allowed the cultivated area to go underreported. Local officials' main motivation in allowing this underreporting seems to have been a desire to grant relief from taxes and production quotas. Against their obligation to deliver grain and other resources to county and provincial governments, local officials seem to have weighed a desire to lighten the burdens on both poor farmers and farmers who were cultivating newly reclaimed lands. The geographic pattern supports this interpretation by showing that the provinces where the most cultivated land is hidden are simultaneously poorer provinces, provinces far from Beijing, and provinces where a greater than average share of lands were recently reclaimed.

The Fall and Rise of Hidden Lands since the 1930s
Now that we have a sense that the cultivated area has been underreported by 38–44 percent around 1985, the next task is to gain a

Table 6.3
Cultivated areas in China by province, 1932–1983

Province	Cultivated areas, usual estimates (thousands of hectares)				Cultivated land correction factor for 1985 (Wang et al.)	Implied share of cultivated land unreported in 1985	Corrected 1983 cultivated area (thousands of hectares)
	1932	1957	1978	1983			
South							
Yunnan	1,748	2,841	2,722	2,833	1.08	0.52	5,893
Guangxi	2,295	2,531	2,563	2,613	0.69	0.41	4,416
Guangdong	2,597	3,827	3,228	3,143	0.81	0.45	5,689
Fujian	1,403	1,489	1,291	1,285	0.31	0.24	1,683
Jiangxi	2,889	2,813	2,523	2,387	0.17	0.15	2,793
Hunan	3,347	3,827	3,437	3,397	0.49	0.33	5,062
Guizhou	1,545	2,033	1,897	1,896	1.62	0.62	4,968
Sichuan	10,363	7,687	6,645	6,525	0.75	0.43	11,419
Hubei	4,300	4,287	3,765	3,699	0.24	0.19	4,587
Anhui	4,875	5,867	4,460	4,440	0.38	0.28	6,127
Zhejiang	2,777	2,252	1,828	1,817	0.47	0.32	2,671
Shanghai	(in Jiangsu)		358	350	0.14	0.12	399
Jiangsu	5,686	6,200	4,649	4,642	0.19	0.16	5,524
North							
Shandong	7,237	9,333	7,291	7,188	0.30	0.23	9,344
Henan	6,532	8,973	7,152	7,079	0.27	0.21	8,990
Shanxi	4,859	4,541	3,914	3,857	0.63	0.39	6,287
Shaanxi	3,039	4,467	3,844	3,749	0.54	0.35	5,773
Gansu	1,745	3,959	3,556	3,552	0.68	0.40	5,967
Ningxia	121	932	894	838	1.31	0.57	1,936

Table 6.3
(continued)

Province	Cultivated areas, usual estimates (thousands of hectares)				Cultivated land correction factor for 1985 (Wang et al.)	Implied share of cultivated land unreported in 1985	Corrected 1983 cultivated area (thousands of hectares)
	1932	1957	1978	1983			
Nei Mongol	1,290	5,543	2,732	5,089	0.39	0.28	7,074
Hebei	8,161	9,002	6,665	6,621	0.14	0.12	7,548
Beijing	(in Hebei)		427	423	0.26	0.21	533
Tianjin	(in Hebei)		467	457	0.38	0.28	631
Liaoning		4,751	4,679	3,665	0.26	0.21	4,618
Jilin		4,719	5,046	4,070	0.34	0.25	5,454
Heilongjiang		7,287	9,168	8,783	0.27	0.21	11,154
Far West							
Qinghai	289	500	596	577	na	na	na
Xinjiang		2,015	3,179	3,158	na	na	na
Tibet			219	229	na	na	na
Totals							
All China	77,096	111,675	99,195	9,8362			142,273
Our China (excluding Qinghai, Xinjiang, Tibet)	76,807	109,160	95,201	9,4398			136,539
South	43,825	45,652	39,366	3,9027			60,830
North	32,982	63,508	55,835	5,5371			75,709
Northwest	6,194	14,902	11,026	1,3228			20,750

Sources: See notes to table 6.1
Notes: 1932 area for Nei Mongol refers to Suiyuan alone. Here Northwest refers to Gansu, Neimongol, Ningxia, and Shaanxi. Ningxia 1957 is 1965 figure from Taiwan's *Atlas of Mainland China*, reported in Dawson 1970, p. 53.

clearer view of the time path of this underreporting. Though this book must leave the task of discerning trends since 1985 to more qualified observers, there are clues to the earlier history of hiding cultivated land from the data and tax collectors. The "alternative cultivated area" series for cultivated area in panel B of table 6.1 offers a best guess on movements since the 1930s.

Cultivated land areas were also underreported back in the 1930s, to a degree that Kang Chao (1970, chapter 8, and appendix tables 1–4) and others have estimated at about 10 percent. Then too the main motivation for underreporting was to avoid taxes. Land area reporting became more complete soon after the revolutionary victory, then fell back during reclamations in the mid-1950s. A major effort to monitor all lands led to virtually full reporting for the year 1957. Thereafter the amount of underreporting began to creep up again. Although we still lack a time series on the percentage underreported, one can imagine that it rose fastest (a) when and where new lands were being reclaimed and (b) when and where the central government had the least control. Such patterns would fit the basic hypothesis that underreporting was primarily a device to avoid land-based taxation (e.g., taxation through quotas). Table 6.1's "alternative cultivated area" series makes only a limited assumption about the timing of the underreporting and evasion: that the extra 38 percent of lands (or 27.5 percent of the true total cultivated area) was already unreported by 1978, and that this remained the share through 1995.[4]

Thus a tentative guess about the share of all cultivated area hidden from the statisticians is that it fell from 10 percent around 1932 to near zero in 1957 and then rose to 27.5 percent by 1978, after which it stabilized.

To set the stage for a later return to the historical geography of China's cultivated land areas, figure 6.1 contrasts the rates of land expansion between 1957 and 1983 implied by the revised figures on cultivated area. Note that in no province did the cultivated area decline by as much as 5 percent, contrary to the widespread fear that it was declining by 15 percent for China as a whole since 1957. Note further that the areas of greatest expansion are those to the west—more remote, more marginal in their yield, and supposedly more prone to erosion. We return to figure 6.1 when discussing desertification later in this chapter.

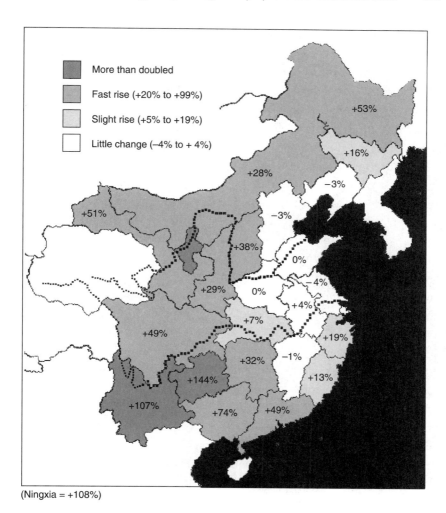

(Ningxia = +108%)

Figure 6.1
Changes in China's cultivated areas, by province, 1957–1983

China's Net Soil Investment since the 1930s

One can now tentatively sketch the time path of China's stock of agricultural land by combining two kinds of educated guesses built into table 6.1: panel A's rough estimates of average topsoil quality and panel B's guesswork about cultivated land areas. The product of these quality and area measures yields a measure of the total endowment of agricultural soil quality, presented in table 6.4. The resulting numbers are to be considered with great care, since we lack a means of placing confidence intervals around them. The main suggestion follows directly from the results just shown in table 6.1. With soil quality changing little and with cultivated area rising, China's overall land endowment—its cultivated area adjusted for changes in soil quality—might have risen by 20–30 percent from the 1950s to the 1980s. Although it may not have risen as much as this, it seems unlikely that the land endowment actually shrank, given the seeming increase in cultivated area and the nondecline in average soil quality.

To be sure, the land endowment did not increase as fast as China's population, as we are reminded by the per capita results in the lower panel of the same table. Yet, as chapter 2 has stressed, such popular calculations of land per capita or food supply per capita have an uncertain meaning in a region or country that has a comparative advantage in selling manufactures to the rest of the world in exchange for imported agricultural product. In an improving and prospering world, China and the rest of eastern Asia could go on reducing their agricultural output or land forever on a per capita basis, as all the other developing countries have done. The point remains, however, that China's total agricultural land area has probably not been declining yet.

Three Lingering Concerns

These rough best guesses about a half century of change in China's agricultural soil endowment may seem to ignore some lingering fears. If the best guesses really imply that China as a whole has gained in cultivated area and has maintained its average soil quality, what happened to the three much publicized dangers of (a) urban encroachment, (b) deser-

Table 6.4
Estimated changes in China's total agricultural land endowment

	1930s level	1950s level	1980s level	Percentage change
In millions of hectares				
China, using quality index from 2 regions	109.5		133.5	21.9
China, using quality index from 8 regions		104.0	133.5	28.3
In millions of hectares per capita				
China, using quality index from 2 regions	0.217		0.131	−39.7
China, using quality index from 8 regions		0.161	0.131	−18.7

Notes: Derived directly from Panels A and B of table 6.1, using the regions listed there. Both the total cultivated area and the total population here refer to all of mainland China, no longer excluding the far west.

tification, and (c) land uses that raise China's vulnerability to floods? The next three sections examine these issues. Table 6.1's estimates already capture two of the three—urbanization and desertification—but they deserve separate consideration nonetheless. The third, land uses that raise flood losses, takes the discussion of the land endowment beyond agriculture.

Soil Losses at the Urban Fringe?
The first lingering question not yet answered explicitly is: Didn't the loss of farm lands to cities, industries, and homes cut the total stock of soil nutrients, even aside from degradation, on any one kind of land that stayed in agriculture? Didn't it remove not only hectares, but *especially fertile* hectares, the ones next to the spreading cities? We turn first to the land area converted and then to urbanization and industrialization's offsetting effects on the fertility of land remaining in agriculture.[5]

The existing estimates of the land agriculture has lost to other uses need to be viewed with caution. A simple example of potential pitfalls in considering the data on land conversion arises from an apparent misstep by a renowned magazine. Consider this recent warning by a not usually pessimistic source: "The bad news is that [China's] crop land is rapidly disappearing. More than 700,000 hectares of cultivated land were taken by construction during the past year." ("Malthus Goes East," 1995, p. 29.) If the publisher of this statement, *The Economist,* is right, then construction takes about 0.53 percent of agricultural land in China a year, or more than 5 percent in a decade. But it appears that *The Economist* used the wrong figure. It seems to be citing the 1994 figure of 708,700 hectares of total decrease in cultivated area. But the adjacent columns in the *China Statistical Yearbook* (China, State Statistical Bureau 1996, p. 355) make it clear that the losses to construction totaled only 245,800 hectares; the rest were "lost" to orchards, harvested ponds, temporary abandonment of flooded lands, and the like. D. Gale Johnson (1995) faults Lester Brown for the same switch of 1994 land areas. The correct figure of 245,800 hectares per annum represents a loss of only 0.19 percent of China's 1985 cultivated area. More generally, over the whole decade 1986–1995, the average annual loss was 198,600 hectares, or only 0.15 percent per annum. Offsetting this loss to construction were the hectares gained by reclamation of other lands for agriculture.

To guard against facile optimism, one must note that the rate of land area loss to construction over the 1986–1995 decade is higher than the rate of loss in earlier decades. To get an idea of what might have been lost to cities and other construction over the half century from the 1930s to the 1980s, it is useful to start from an estimate in a 1984 agricultural atlas. According to the atlas, the land cumulatively taken by cities, towns, villages, and roads in 1983 was about 4.46 percent of the land in China (excluding the desert far West and the islands). Fixing the proportion between nonagricultural population and nonagricultural land use implies that agriculture lost 2.16 percent of its land area to construction between about 1933 and 1983, or an annual compound rate of 0.04 percent. So to decide on the importance of the of land conversions from agricultural to "urban" uses, one can start from these three baseline facts:

1. In the most recent decade (1986–1995), construction took 0.15 percent of the cultivated area per year, or 1.5 percent per decade.

2. Over the preceding five decades, construction took about 0.04 percent a year, or 0.4 percent per decade.

3. The rate of loss to cities and industry is therefore rising.

The next step in appraising the urbanization issue is to confront the suspicion that the lands being lost at cities' edges are more productive than the average lands remaining in agriculture, so that the rate of accelerating loss may be greater than mere hectare numbers can reveal. This appears to have been true. Related work has estimated that the land thus lost was somewhat richer than the average land.[6] That 1933–1983 loss of 2.16 percent on the edge of cities and villages lost China's agriculture soil nutrients as shown in table 6.5. To sound the loudest alarm possible, one could take the largest percentage losses as reflective of the soil productivity loss of this half century, suggesting a loss of up to 3.53 percent of nutrients in the north or up to 2.31 percent in the south. These are less than revolutionary, however, when spread over a whole half century.

There is an important offset to these limited losses of better than average lands at the urban border. The buildup of nonagricultural population, especially around urban centers, tends to raise the levels of organic carbon and organic nitrogen in local agricultural soils. As observers have noted since the 1930s, China's urban-industrial expansion has raised the

Table 6.5
Estimated losses in soil nutrients per hectare due to population growth and the shift from agriculture, China, 1933–1983

	North China	South China
Soil organic matter	−2.75%	−2.31%
Soil nitrogen	−2.19%	−1.99%
Soil phosporus	−3.53%	−2.28%
Soil potassium	−2.16%	−1.91%

Note: Minus signs indicate a loss of soil nutrients.

supply of manures, and to a lesser extent chemical fertilizers, for farms nearest to the cities and the industrial activity. At the same time, the proximity of a greater nonfarm population also raises the marketability—or, under Chairman Mao, the planning urgency—of supplying that population with perishable foods at short distance.

Figure 6.2 sketches the possibly offsetting effects of urbanization and industrialization on the nutrients of nearby soils. The figure attempts to show that the positive effect of a greater density of nonagricultural population could be greater than, or less than, the loss of soil nutrients in the areas taken away from agriculture. The well-known losses just roughly quantified are represented by the area marked with a minus sign in the figure. Against these should be weighed those possible positive effects on the fertility of the soils near the spreading cities, represented by the area with the plus sign. Which is greater?

Related calculations have imagined how the magnitudes might have worked out for China over the half century ending in 1983 by borrowing regression coefficients from the soil history samples (Lindert, Lu, and Wu 1996a, p. 1177). The positive effect of growth in the nonagricultural population appears to have been strong enough to outweigh the direct losses of farmland in terms of soil organic matter and nitrogen. The effects of rising agricultural population were more negative. Overall, urbanization and industrialization and population growth together have augmented the agricultural stocks of soil organic matter and total nitrogen, though they have cut the endowments of soil phosphorus and potassium. The results, in other words, seem to land on both sides of zero, with a net effect somewhere near zero.

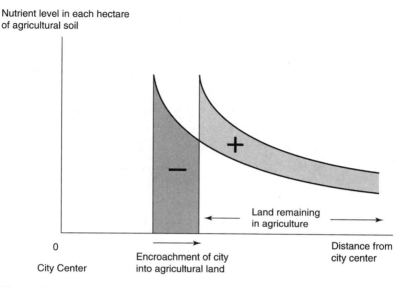

Figure 6.2
Stylized picture of the effects of urban encroachment on the total soil-nutrient endowment

Urbanization and industrialization thus may have offsetting effects on agriculture's land endowment. To resolve whether the net effect has been to raise or lower that endowment over recent decades and what that net effect will be over upcoming decades, future research must quantify these different effects more extensively than has been possible here.

Desertification

[B]y AD 2000 there may be roughly twice the northern desert area as there was in 1949, despite China's courageous efforts to build a great green wall.
—K. Forestier

At the desert fringe, as at the urban fringe, one may still wonder what happened to the fully erosion and advancing sand dunes decried by the literature cited in chapter 1. If the desert continues to encroach, are humans being forced to maintain their cultivated area by switching to increasingly marginal lands somewhere away from the advancing desert?

It is time to reconcile the national aggregates with the long-standing perception of a danger of desertification. The most direct kind of evidence is obtained by supplementing the national figures with estimates of cultivated area and soil quality in the provinces of the long desert border. There is a clear conflict in evidence to reconcile here, however. Against the anecdotal and photographic evidence of land loss in the Great Bend of the Huang He and in the vast borderlands where sand meets grass we have those provincial figures showing *expanding* areas under cultivation even in the classic desert-prone provinces. We already saw a hint of this in figure 6.1, which shows, it seems, that cultivation between 1957 and 1983 expanded even faster than average in the four desert-prone provinces of Gansu, Ningxia, Shaanxi, and Nei Mongol.[7]

If farmers are really working greater and greater land areas in the desert-prone northwest, as figure 6.1 suggests, how could past authors have missed this trend? Part of the answer may have been their preference for visual and narrative testimony over quantification. More to the point, however, the numbers available to the authors of the literature before the 1990s featured the official numbers on cultivated area, which missed the extra cultivated land discovered by satellite surveillance and now accepted by the official statisticians themselves. Table 6.6 outlines the difference that the new corrections make for those four provinces in the desert battle zone in the 1957–1983 period. It is understandable that without the recent corrections one could see no contradiction between the available numbers and the visual signs and narratives that indicated that farmers were forced to retreat in the face of attacks by wind, water, and sand. Yet the corrected figures on cultivated area deny the retreat. The rough estimates of soil quality agree, showing no net decline in quality for the agricultural regions of the northwest. So do the data on topsoil thickness, which show no thinning of the soils of the northwest between the 1950s and the 1980s.[8]

It appears that the main cause of the continuing struggle against the desert is probably *not* that the desert has advanced into the farms of China. On the contrary, the struggle continues mainly because the farms of China have advanced into the desert. This result should not be surprising given the intensity of China's revegetation and reclamation efforts in the northwest. In the future the farmlands may retreat for a reason quite

Table 6.6
Percentage change in cultivated area in four provinces, 1957 to 1983

	Implied in table 6.3 by	
	The official series	The corrected series
Gansu	down 16%	up 29%
Ningxia	down 10%	up 108%
Shaanxi (including the Great Bend of the Huang He)	down 10%	up 51%
Nei Mongol	down 8%	up 28%

different from any defeat inflicted by the desert. Rather, cultivation may retreat in favor of more grasslands and reforestation, as China's labor force shifts to the east to earn higher incomes in the rising urban industrial centers.

Lessons from Floods

A final land use issue not sufficiently addressed by national figures on soil quality or cultivated area transcends this book's topic of soil and agriculture. The productivity of any given land use should be measured not only by what it yields in normal times and how it conserves or degrades the soil, but also by how it affects the severity of rare disasters. For millennia China has faced climatic disasters in the form of floods, droughts, and high winds. Some forms of land use mitigate a natural disaster, whereas others worsen it. Any treatment of soil history that suggests broad conclusions about the effects of human management of the land should also note the disaster-mitigating or disaster-worsening effects of the same land management, even if the land use issue in question is not strictly a question of trends in the soil endowment.

A case in point is the series of serious floods that took thousands of lives and destroyed homes along the Yangzi and in northeast China in 1998. Although the disaster was caused primarily by an extreme climatic event, most commentators agree that land use practices added to the destruction. In particular, the floods seems to have been made worse (1) by deforestation of the far upper reaches of the Yangzi, and (2) by the gradual conversion of lakes and wetlands into levee-protected farmlands downstream. Of the two land uses, the upstream deforestation seems to be less important than the downstream loss of lakes and wetlands. The upper Yangzi has not been heavily forested for centuries. Though reforestation might help to hold some extra rainfall in a torrent and help limit the siltation of reservoirs such as the one being created by the Three Gorges Dam, it is generally thought that deforestation—at least, twentieth-century deforestation—in the upper Yangzi does not have a great role in flooding. In the floods of 1998, most of the torrent fell in catchment area that had little forest area even 100 years ago. More serious is the loss of a wetland reserve in the middle and lower Yangzi and in the wetlands of the northeast. About one third of the lake and wetland area of the middle and lower Yangzi has been developed as farmland since 1949, especially in Hubei and Hunan near Wuhan. This has compromised water storage, so that the floods of 1998 were harder to alleviate.

The issue of flood control relates to this chapter's history of the agricultural land endowment both in its comment on past land use and its suggestions about future land use. Part of the extra flood danger from land use has been created by excessive conversion in the past of wetland reserves to farmlands protected by levees, particularly in the eras of all-out pushes to grow more grain. Perhaps in the near future plans will be made to reconfigure the levee system so that a large designated agricultural area could yield to flood when necessary, allowing other areas to escape being flooded. This will be easier to do the more China shifts both labor and land away from staple agriculture, freeing up some lands as water control reserves. At the present, the re-creation of wetland reserves is not easy, since about 20 million people live in the filled-in former wetlands.

Summary

Several tentative findings offer new perspectives on the prospects for agriculture's soil endowment in a growing China. However, reminders about the quality of the underlying data must precede any summary of these findings. Although the available soil data are abundant for the 1980s, far fewer soil profiles are available for the 1930s and 1950s. The agricultural data for counties in 1985, on which the regressions are based, contain their own errors. Some errors have been weeded out during data processing, but some doubtless remain. In particular, the misreporting that has concealed over a quarter of China's cultivated area from official statisticians as of the mid-1990s may already have complicated the patterns in the 1985 data used here. The only cure for these dangers is more cross testing and more data.

Fresh clues about long-debated trends invite further exploration. One such clue is that over a recent half century China's soil quality apparently slipped in some respects, but not overall. Interestingly, the negative trends for some regions and time periods do not correspond to popular fears. The soil trends are probably not worse in the erosion-prone parts of China than elsewhere, and probably not worse since the 1950s than earlier. Weighing the mixture of trends in different soil characteristics suggests that the average quality of China's cultivated soils rose modestly from the 1950s to the 1980s.

Exploring the soil-agriculture nexus in chapter 5 yielded measures of the marginal productivities of different soil characteristics, allowing this chapter to combine these marginal productivities with the historical changes in the soil in order to suggest net changes in China's aggregate agricultural land endowment. It appears that soil productivity is indeed improved by reducing alkalinity in the north, by reducing acidity in the south, and by raising total soil potassium in the potassium-poor south. The effect of raising soil organic matter and soil nitrogen in the same proportions is about zero.

These estimates of the productivity effects of soil characteristics suggest that overall soil productivity did not change significantly in north China between the 1950s and the 1980s. Soil productivity in the south might have risen by 7.4 percent over the same era, but some doubts about the underlying figures calls for caution. The safest conclusion about the overall trend in China's soil quality is that it at least did not decline between the 1950s and the 1980s.

Against this background of stable soil quality, the land areas cultivated by China's farmers increased somewhere between 20 and 30 percent between the 1950s and the 1980s. This finding is based on the satellite-aided revision of the estimates of cultivated area in the 1980s. This revision has had a considerable impact on our impressions about China's total agricultural land base, especially for some western provinces. Combining the nondecline of soil quality with the 20–30 percent increase in cultivated area suggests a definite expansion of China's overall endowment of agricultural land.

Losses of agricultural land to cities, industry, and construction in general appear to be smaller than others have previously suggested. The land areas lost per year are not yet large, despite journalism to the contrary. Furthermore, the growth of cities and of the nonagricultural economy in general may actually raise the productivity of the lands still left in agriculture. Some clues for China between the 1930s and the 1980s suggest that this may indeed have happened.

The desertification issue also needs some reinterpreting in the light of the evidence presented here. Aside from anecdotal evidence, the main quantitative data on the struggle against the desert are those estimates of the cultivated area available in Gansu, Ningxia, Shaanxi, and Nei Mongol. Now that these estimates have been corrected upward for the 1980s and 1990s, it appears that the desert has not advanced into the farms of China so much as the farms of China have advanced into the desert, without any clear loss in the average agricultural productivity of those northwestern farmlands.

III
Indonesia

7

A Half Century of Soil Change in Indonesia

Fortunately, it is now possible, thanks to a data set developed in the mid-1990s with the cooperation of the CSAR, to estimate how Indonesia's soils have changed over the half century 1940–1990. This chapter opens up the first quantitative soil history of Indonesia. The changes it traces in the Java-Madura core are very different from those it finds in the other islands of the archipelago.

Many of the soil trends on the crowded and long-settled islands of Java and Madura resemble those for crowded and long-settled eastern China, especially the eastern provinces of subtropical south China. Both settings had a period of noticeable decline in soil organic matter and nitrogen. That period was 1940–1970 in Java[1] and 1950s–1980s in south China. Both settings also had a general buildup of total soil phosphorus and total soil potassium. And in both Java and south China, an early successful abatement of soil acidity was followed by a slight return toward acid conditions, with the respective turning points again being sometime around 1970 and around the 1950s.

Trends on the settlement frontier of Indonesia's outer islands[2] were very different from those in Java-Madura, and offer a unique historical glimpse of soil dynamics starting with soil samples taken in never-cultivated virgin forests and wastelands that have no Chinese counterpart in the twentieth century. The contrast between the outer islands and Java offers a useful further clue about human impacts on the soil, one that supports our main conclusions about soil chemistry trends in the cultivated areas of both countries.

Preparing the Raw Materials

The Indonesian historical soil sample of the 4,562 best-detailed profiles gathered at the CSAR in Bogor covers time, space, and land use types unevenly. As shown in figure 7.1, waves of heavy soil sampling occurred both before and after Indonesia's independence, shaped partly by political history. The best-covered period before 1985 resulted from the great colonial wave of soil survey work in 1938–1941, which was cut short by the Japanese occupation. The postwar waves covered here are 1953–1962, 1970–1973, and the 1980s. Adding the post-1985 computer data now available at the CSAR would make the present ongoing wave of soil sampling as great as the 1938–1941 wave. Throughout this history, the soil surveying rotated among the islands. Java received a little more than half the coverage consistently to about 1970, and has received less than half the attention since.

The Indonesian data set allows us to study soil conditions under five different categories of land use separately. As shown in table 7.1, these five categories are fairly distinct, despite some overlap. Paddy (*sawah*) fields are invariably flat and cropped at least once a year. They tend to be irrigated, or even flooded, and tend to involve rice. The other category of annual field crops, called *tegalan* in the original data, has fuzzy edges. It usually has each of the following characteristics: upland, rain-fed, sloping, tilled, and dominated by nonrice food crops (*palawija*), though it does not fit any one of these characteristics neatly. For example, *tegalan* is sometimes irrigated, and its crops include such cash crops as tobacco and sugar. The third category, tree crops, consists of land used for crops such as rubber, coffee, tea, tree fruits, palm oil, teak, and cultivated bamboo. The fourth category, "fallow," covers lands lightly or previously cultivated by humans. Only a small share of it is under shifting or slash-and-burn cultivation (*ladang*), both in the sample and in Indonesia.[3] More common are those infamous usually posthuman vegetation types, the invasive imperata grass (*alang alang*) and unmanaged bush (*belukar*). Recently, however, the "fallow" category has come to consist increasingly of useful grasses and green waste, which affects a few results below. The fifth category, "primary forest (*hutan primer*) and waste," presents a useful baseline against which to judge the total human impacts on the soil.

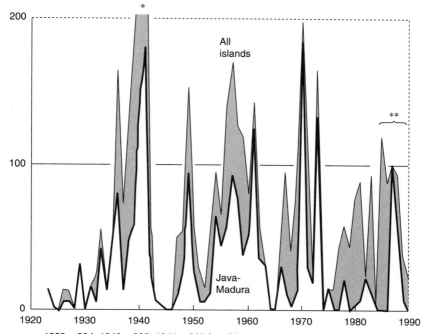

* 1939 = 204, 1940 = 238, 1941 = 305 for all Indonesia
** plus hundreds more profiles form 1985 on,
   in the CSAR computer files not available
   for this study.

Figure 7.1
Numbers of soil profiles used in the Indonesian sample by year, 1923–1990

Table 7.1 shows that lands in the five categories are not evenly spread over time and space. Before 1936, most of the data fall into the category "no land use given." For Java, we can follow the four human-affected land types fairly well from 1938 onward, but few profile data are available from primary forest soils in Java. For the outer islands, there are more samples from forest soils, especially in 1974–1986. The outer-island soils under the categories "tree crops" and "fallow" were frequently sampled from 1938 onward, but the tilled-field categories (paddy and *tegalan*) were not.

A reminder about the land use data is in order. Although the data are often rich in detail about land use, they are vague in two ways. First, some seem to refer to the vegetation on a whole set of fields, not just the site

Table 7.1
Numbers of soil profiles used for Indonesia, by broad land-use category and era, 1923–1990

| | Broad land-use category | | | | | | |
	Paddy (sawah)	Upland field crops (tegalan)	Tree crops	Fallow (grass and bush)	Primary forest and waste	No land uses given	Totals
Java-Madura							
1923–1935					4	180	184
1936–1944	45	67	51	41	12	351	567
1947–1957	25	114	108	61	13	121	442
1958–1967	45	90	140	69	4	54	402
1968–1973	37	281	31	30		3	382
1974–1986	20	22	23	11	6	1	83
1987–1990	45	51	48	21	4	1	170
Totals	217	625	401	233	43	711	2,230
Outer islands							
1926–1937						192	192
1938–1942	2	11	80	111	14	265	483
1947–1957	3	10	129	157	22	107	428
1958–1967	30	4	102	150	15	8	309
1968–1973	27	1	151	61	7	10	257
1974–1986	11	8	201	233	113	16	582
1987–1990	11	16	20	21	13		81
Totals	84	50	683	733	184	598	2,332

where the profile was taken. Second, as chapter 3 warned, the data do not trace the history of land use at the individual sites. They do not tell us how long a particular site has had its current land use, which complicates the task of interpreting human impacts. To further illustrate a point made in chapter 3, suppose that continuous *tegalan* tillage, under average practice, reduces the nitrogen content of a given field linearly from 0.2 percent to 0.1 percent over five years, after which seven years of lying fallow returns the plot linearly to 0.2 percent nitrogen. Soil samples taken evenly over five years of *tegalan* and then over seven years of fallow might give the impression that tilling had no effect on nitrogen, relative to fallow, since the average nitrogen was 0.15 percent both over the tillage years and over the fallow years. So we must judge all results in the light of what is generally known about the usual dynamics of soil conditions from experience elsewhere.

Revealing the Patterns Statistically

The journey from a historical sample of soil profiles to a judgment of soil trends is a long one, because soils are so heterogeneous across a country and even across plots on a single farm. Any statement about how humans have changed the soil over large areas must be one that discusses how the soil has changed in environments where all the nonhuman determinants of soil quality have been held constant over time. The research strategy must also control for any systematic bias in the way in which the data were collected.

The Generic Equation Revisited

To confront these issues, let us first return to chapter 3's overall statistical strategy, then survey the myriad control variables needed to implement that strategy in the Indonesian setting. The strategy begins with chapter 3's generic regression equation

$$Y = a + bH + cX + dS + e,$$

where

Y is a vector of the dependent variables we seek to explain, namely ln (organic matter), ln (N), ln (P_2O_5), ln (K_2O), acidity, and shallowness of the topsoil layer;

a, *b*, *c*, and *d* are vectors of coefficients;

H is a vector of historical trend variables, differentiated by land use category, as explained below;

X is a vector of slow to change nonhuman attributes of the site (soil texture, terrain, parent materials, soil classes, precipitation, and other landscape features);

S is a vector of systematic biases in some of the data sources, varying over time and space; and

e is an error term.

Data availability dictates the list of dependent variables (the *Y*s). Over many decades these six attributes, plus soil texture, are the ones most consistently measured.

The list of *X*'s begins with five physical influences on soil chemistry: climate, terrain, soil texture, soil parent material, and USDA soil class. Climate is less a function of temperature differences in Indonesia than in China. Its variations are more easily captured through the local precipitation patterns.[4] Figure 7.2 summarizes the key precipitation variable for Indonesia by mapping climate zones according to their average number of wet months minus their average number of dry months.[5] As is readily evident, the western half of the area is much wetter than the eastern half (other than Irian Jaya). Even Java is split between a wetter west and a drier east. Experts have long known that in the tropics a wetter climate depletes potassium and raises acidity and has more nuanced effects on other nutrients. In the absence of sufficient irrigation infrastructure, a wetter climate also imposes less constraint on the number of crops that can be raised per year. This chapter confirms the expected effect of the wetter climate on soil potassium and acidity, and chapter 8 confirms its effect on agricultural output.

The remaining physical influences play the same roles they played for China in part II. Sloping terrain suffers depletion of soil potassium and higher acidity if it is not terraced. The effects of soil texture are potentially nonlinear and are captured here by quadratic functions of the shares of sand and clay in the topsoil. The list of physical determinants also includes binary variables representing nine parent materials and more than a dozen soil classes, as mapped by CSAR.

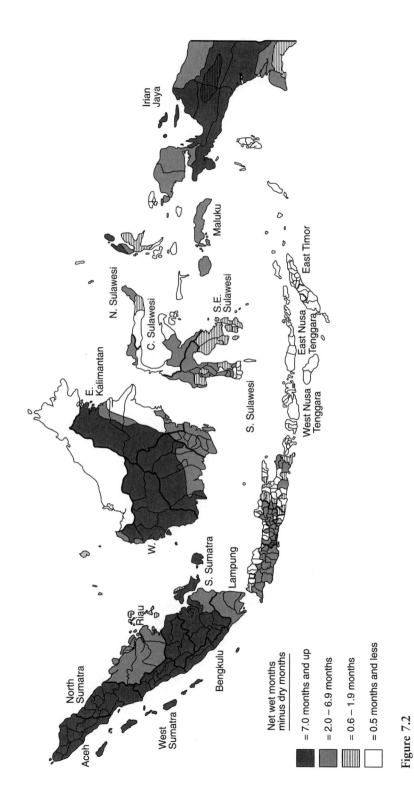

Net wet months
minus dry months

■ = 7.0 months and up

▨ = 2.0 – 6.9 months

▥ = 0.6 – 1.9 months

□ = 0.5 months and less

Figure 7.2
Net wet months per year, based on CSAR 1983 agro-climatic map

Time Trends as Human Fingerprints

The crucial task to accomplish in gathering data on independent variables that explain soil conditions is to separate human from nonhuman effects. Fortunately, the ability to distinguish human versus nonhuman effects is largely a matter of time versus space here. A sample spanning a few decades can distinguish human impacts from other effects. It is long enough for human management to show a clear effect, yet short enough not to be subject to most nonhuman effects, even climate and volcanic activity. Hence the historical time trends summarized by H in the equation can be attributed to human influences—as long as the equation is correctly specified. The spatial X variables can be thought of as non-human, since even spatial differences in human treatment of the land are largely responses to nonhuman physical variables.

The time effects (H), which play the central role here, involve interactions between the type of land use and the passage of time. Calculations are made of the interactions of each of the six land use categories (again, *sawah, tegalan,* tree crops, fallow, primary forest, and no land use given) with each of these six *time slope variables:*

• the number of years before 1940 (which equals 17 for 1923, zero for 1940 and beyond, etc.);

• the number of years after 1946 (= zero up through 1946, and values up to 44 for 1990); and

• the numbers of years after 1955, after 1960, after 1970, and after 1980.

The staggered start-up of extra slopes for later years is a spline function way of allowing for switches in trend without unrealistic discontinuities at any year. The choice of turning point years was based on a prior search for years that showed a significant change in trend for at least one soil characteristic. The "winning" transition years were 1940, 1946, 1955, 1960, 1970, and 1990.[6] Not all of these thirty-six time and land use terms were used, however, since some of the combinations involved too few data points. The number of combinations varied from fourteen to twenty-six across the soil attribute equations and the two parts of Indonesia. Two other time effects (postwar shift and the period 1974–1986) were added as S variables, to which we turn in the next section.[7]

The same procedures that turn up plausible and useful changes in long-run trend can also reveal anomalies. In what follows, there is a recurring anomaly in the contrasts of soil values "predicted" by the regression for 1955 and for 1960. In many cases, either 1955 or 1960 values can stick out of the longer time path in a way that defies explanation. The most likely answer for this anomaly is that attempts to purge special influences of the places sampled in different periods were unsuccessful. Although simply averaging the 1955 and 1960 values could smooth the way to all the conclusions suggested here, it seemed better to leave the 1955 versus 1960 anomalies in plain view.

In Search of Source Bias

As chapter 3 warned, any use of several decades of soil profile data must ensure that changes in measurement technique or survey selectivity are not mistaken for changes in the true condition of the soil. Did the soil survey teams really measure things the same way over the years? CSAR manuals have given the same instructions throughout the postwar period, and the detailed soil profile forms filled out since 1938 have always called for the same attributes. For example, pH is consistently reported in H_2O suspension and P_2O_5 and K_2O are always available in a 25 percent HCl measure. One change is easily corrected for: In mid-1957, there was a switch from reporting organic matter (*bahan organa*) to reporting organic carbon (*C organa*), which calls for multiplying the post-1957 values by 1.7 to ensure comparability. Yet even with uniform reporting, changes in laboratory procedures could have artificially raised or lowered the attribute values. Although CSAR officials know of no such changes, the possibility that they occurred should not be ruled out.

Indeed, the regression results themselves suggest that such shifts in procedures might have occurred. Specifically, in the reporting of N, P_2O_5, and K_2O, the values for 1974–1986 are usually one third to one half lower than any other characteristics can explain, regardless of where in Indonesia the profile was taken (yet no such shift occurred for organic matter or pH). The drop in those three attribute values started so abruptly between 1973 and 1975, and ended so abruptly between 1985 and 1987, that it cannot reflect any real soil change. In those years most samples were taken in the outer islands and were, reportedly, shipped to the same central laboratory in Bogor. Although the 1974 drop and the 1987 rise cannot be considered proof of shifts in otherwise consistent laboratory procedures, we need to allow generously for this possibility when judging trends. In all that follows, the regression procedure assumes that any estimates for the 1980 benchmark fall exactly on the trend line from 1970 to 1990, and that the observed values of N, P_2O_5, and K_2O were lower in 1980 solely because of a temporary measurement bias. Yet any interpretation must allow that the 1980 values could have been truly lower.

Survey selectivity, like any change in a measurement technique, is another serious threat spelled out in chapter 3. Surveyors usually do not make equiprobable random draws from all the possible sites. Rather, they sample soils for a particular purpose, and that purpose may change over time and space in ways that can mislead. We have just seen one reason to fear such a bias for Indonesia: Perhaps the anomaly of low nitrogen, phosphorus, and potassium in 1974–1986 stemmed not from true soil changes (or from laboratory changes) but from an implicit policy of turning attention to lower-productivity soils in the outer islands. Perhaps the X variables that were supposed to control differences in site cannot pick up the real selection biases.

To search for biases that the H and X variables cannot explain, I looked for places that had strangely high or low values of soil characteristics. I concentrated on anomalous differences across administrative boundaries: on differences between whole islands and on departures of individual districts from the soil characteristics the Xs and Ts could explain in early regression runs. Early regressions suggested that on Java there might be fixed-place biases associated with Jakarta, Surabaya, Bali, and thirteen rural districts (*kabupaten*). For the outer islands I have explored possible fixed-place biases associated with four districts in Sumatra and with each island relative to Sumatra (dividing Sulawesi into two zones with different behavior).

The Estimated Soil Equations

The resulting statistical regressions for each soil characteristic are long, involving from seventy-six to ninety-three independent variables (Hs, Xs, and Ss). Table 7.2 illustrates, with a single equation shown in partial detail, the equation for the log of organic matter in Java, drawing on a total of 1,777 profiles from 1923 to 1990. Table 7.3 adds a summary inventory of all the soil chemistry equations, backed up by Appendix D's listing the full set of ten equations.

The first set of coefficients in table 7.2's regression shows that organic matter tends to be higher in land for other uses than in paddy soils. Although the individual land use coefficients appear insignificant, they are significant as a group vis-à-vis paddy soils. Similarly, the individual time slopes representing switches in trend appear statistically insignificant when viewed separately in table 7.2, though they add up to significant cumulative trends.

Before turning to the central question of trends, let us note that the fuller versions of the regressions confirm some plausible place-to-place determinants of soil chemistry. First, the role of soil texture follows some consistent patterns. If Indonesia's soil endowment has been less sandy (than the 18 percent sand on Java and 32 percent sand at the average outer-island site), it could have maintained more nitrogen, phosphorus, and potassium, more organic matter, and (for the outer islands) more pH neutrality. Less clay would also have helped in every respect, except that more clay could have retained more organic matter and nitrogen on the outer islands.

Climate also matters, as expected. In general, the wetter areas have higher levels of organic matter, higher nitrogen and phosphorus, lower potassium, and greater acidity, especially within Java, though the pattern is not perfectly monotonic. Terrain also matters, though less clearly than soil texture or climate, other things being equal. Flatter terrain seems to be associated with higher levels of phosphorus and potassium and, in Java, with pH closer to neutrality.

Soil Trends on Java

Since each of the time slopes, like the ones illustrated in table 7.2, measures a change in trend, they can be added together to test the direction, speed, and significance of cumulative trends since 1940. For the year 1973, say, the cumulative trend equals the postwar shift term plus (27 · the coefficient on the number of years since 1946) plus (18 · the

Table 7.2
A single regression in partial-full detail: Determinants of organic matter in Java topsoils, 1923–1990

Independent variables	Dependent variable (ln of % of organic matter)	
	Coeff.	$|t|$
Fixed effects of land use type (relative to paddy)		
Tegalan (nonpaddy field crops)	0.1482	(0.56)
Tree crops	0.2379	(1.24)
Fallow (grasses, bush, ladang)	0.1990	(0.93)
Primary forest	0.3476	(1.48)
No land use given	0.2727	(1.57)
Spline function time slopes by land use type (for cumulative trend results, see table 7.5)		
Post–World War II, any use	−0.1107	(1.07)
Data 1974–1986, any use	0.0984	(1.11)
Paddy, number of years since 1947	−0.0181	(0.57)
Paddy, number of years since 1955	0.0182	(0.25)
Paddy, number of years since 1960	−0.0155	(0.26)
Paddy, number of years since 1970	0.0203	(0.95)
Tegalan, number of years since 1947	−0.0390	(1.43)
Tegalan, number of years since 1955	0.0489	(1.12)
Tegalan, number of years since 1960	−0.0191	(0.61)
Tegalan, number of years since 1970	0.0060	(0.46)
Tree crops, number of years since 1947	−0.0135	(0.73)
Tree crops, number of years since 1955	−0.0047	(0.12)
Tree crops, number of years since 1960	−0.0060	(0.18)
Tree crops, number of years since 1970	0.0369	(1.72)[b]
Fallow, number of years since 1947	−0.0234	(1.00)
Fallow, number of years since 1955	0.0291	(0.57)
Fallow, number of years since 1960	−0.0315	(0.73)
Fallow, number of years since 1970	0.0354	(1.54)
Primary forest, number of years since 1947	−0.0859	(2.52)*
Primary forest, number of years since 1955	0.3734	(3.54)**
Primary forest, number of years since 1960	−0.1799	(1.37)
Primary forest, number of years since 1970	−0.2258	(1.89)[a]
No land use given, number of years pre-1940	0.0121	(1.14)
No land use given, number of years since 1947	−0.0387	(1.55)
No land use given, number of years since 1955	0.0409	(0.64)
No land use given, number of years since 1960	−0.0801	(1.22)
No land use given, number of years since 1970	0.1234	(2.18)*

Table 7.2
(continued)

Independent variables	Dependent variable (ln of % of organic matter)	
	Coeff.	$|t|$
Effects of texture and topsoil depth		
Sand %	0.0361	(4.66)**
Sand % squared	−0.0005	(7.49)**
Clay %	−0.0042	(0.61)
Clay % squared	0.0001	(1.04)
Sand % times clay %	−0.0006	(4.54)**
No texture given	0.0990	(0.43)
Shallowness of topsoil	0.9793	(3.84)**
Numbers of additional variables		
• Terrain	5	
• Parent material (geology)	9	
• Precipitation	7	
• Soil classes	17	
• Districts with peculiar fixed-place effects	16	
Total number of independent variables	93	
• Constant term		
Adjusted *R*-squared	0.404	
Dependent variable mean	0.960	
Standard error of estimate	0.543	
Regression type	OLS	
Number of observations	1,777	

Notes: Statistical significance of coefficients (two-tailed): **1-percent level, *5-percent level, [a]6-percent level, [b]7-percent level.

As partially implied by the headings over independent variables, the base case to which others are compared is an alluvial tropaquept in flat lowland paddy with 5–6 months wet and 2–3 months dry around the year 1940. Such a soil profile might have been taken near Semerang in Central Java. The base case for the outer island regressions has the same attributes except that it has 10–12 wet months and 0–1 dry month. Such a profile might have been taken near Kisaren in North Sumatra.

A "wet" month averages more than 200 mm of precipitation. A "dry" month averages less than 100 mm.

Shallowness of topsoil = 1/(the median depth of the topsoil layer) = 2/(depth at its top + depth at its bottom).

Table 7.3
An inventory of regression equations for the determinants of soil chemistry characteristics in Indonesia, 1923–1990

	Dependent variables and equation numbers									
	(ln of % of) organic matter		(ln of % of) total nitrogen		(ln of mg/100g of) total P_2O_5		(ln of mg/100g of) total K_2O		max (6 – pH, 0) acidity	
	(1) Java	(2) Outer islands	(3) Java	(4) Outer islands	(5) Java	(6) Outer islands	(7) Java	(8) Outer islands	(9) Java	(10) Outer islands
Independent variables										
Time effects on different kinds of lands (for cumulative trend results, see table 7.4)										
Trends in paddy soils	4	1	4	1	5	4	5	4	4	4
Trends in *tegalan* soils	4	1	4	1	4	3	4	3	4	2
Trends in tree crop soils	4	4	4	4	5	5	5	5	4	4
Trends in fallow lands	4	4	4	4	5	4	5	4	4	4
Trends in primary forests	4	1	3	1	3	1	3	1	4	1
Trends in soils of unknown use	5	3	4	3	4	5	4	5	5	5
Shift terms: Postwar, 1974–1986	2	2	2	2	2	2	2	2	2	2
Numbers of other independent variables in the same equations										
• Fixed effects for land use type relative to paddy (cf. table 7.2)	5	5	5	5	5	5	5	5	5	5
• Topsoil texture, depth	7	7	7	7	7	7	7	7	7	7
• Terrain	5	5	5	5	5	5	5	5	5	5
• Parent material (geology)	9	12	9	12	9	12	9	12	9	12
• Precipitation	7	9	7	9	7	9	7	9	7	9
• Soil classes	7	4	7	4	7	4	7	4	7	4

Table 7.3
(continued)

| | Dependent variables and equation numbers | | | | | | | | | |
| | (ln of % of) organic matter | | (ln of % of) total nitrogen | | (ln of mg/100g of) total P_2O_5 | | (ln of mg/100g of) total K_2O | | max $(6 - pH, 0)$ acidity | |
	(1) Java	(2) Outer islands	(3) Java	(4) Outer islands	(5) Java	(6) Outer islands	(7) Java	(8) Outer islands	(9) Java	(10) Outer islands
• Districts with peculiar fixed-place effects	16	8	15	8	13	8	13	10	13	6
Total number of independent variables	93	76	90	76	91	84	91	86	90	80
• Constant term										
Adjusted R-squared	0.4036	0.2822	0.4173	0.2839	0.2808	0.4896	0.3760	0.5253	0.4267	0.4212
Dependent variable mean	0.9602	1.3927	-1.9991	-1.6590	3.8130	3.2548	3.5709	3.2797	0.1583	0.6901
Standard error of estimate	0.5428	0.6594	0.4742	0.5942	0.7799	0.7686	0.7092	0.7750	0.6716	0.7122
Regression type	OLS	OLS	OLS	OLS	OLS	OLS	OLS	OLS	tobit	tobit
Number of observations	1,777	1,703	1,302	1,613	2,115	2,176	2,126	2,240	2,191	2,312

Notes: See notes to table 7.2.

Cell values = numbers of independent variables used.

For the two tobit equations, the dependent variable is Acid = max $(6 - pH, 0)$. Its R-squared is the squared correlation between observed and expected values. The share of observations with acidity above zero is .4135 for Java and .7011 for the outer islands. Multiplying by this decimal fraction converts each regression coefficient into a predicted overall effect at mean values of the independent variables. On the tobit regression approach to dealing with bounded dependent variables (e.g., bounded by zero, as here), see Maddala 1983.

coefficient on the number of years since 1955) plus (13 · the coefficient on the number of years since 1960) plus (3 · the coefficient on the number of years since 1970). Some resulting cumulative trends are given in figure 7.3 and table 7.4. The absolute level of each soil attribute is the actual average value for a mix of places sampled in 1958–1962, adjusted for other dates according to the cumulative trends.

Note that the set of dates for which estimates are ventured here is more limited than the long 1923–1990 time span of the data set used in the regressions. The dates shown are only those around which there was a sufficiently abundant and diverse sampling of soils over a five-year period to warrant an estimate. In general, estimates are not shown for any benchmark year in which five years of data turned up less than fifteen sample observations from fewer than five districts.

For long-crowded Java, we can follow a half century of trends in most attributes for all the land uses reflecting current or recent human cultivation. The organic matter of Java's topsoils seems to have dropped from about 1940 to about 1970 and to have risen thereafter, or so one infers from the soils sampled under tree crops and under those mostly postcultivation fallow fields consisting of pasture, imperata grass, bush, and shifting cultivation (*ladang*). For the main field crop categories, paddy and *tegalan*, 1970 again looks like a low point, though sampling before about 1955 is too thin to establish the 1940–1955 trend. The tentative verdict on organic matter in Java is that it has decreased since 1940, with a large decline to 1970 and a lesser rise thereafter. One would expect roughly similar trends for total nitrogen, given that carbon to nitrogen ratios tend to equilibrate over the long run. Trends are harder to judge for nitrogen than for other soil nutrients, however, because it was less faithfully recorded than other attributes before 1955. If it is true that organic matter and nitrogen both declined from 1940 to 1990, as the organic-matter data suggest, then Java's experience was similar to that of eastern China between the 1930s and the 1980s (chapter 4).

Over the same long period, the clearest nutrient gain in Java was that for phosphorus. The estimates imply that over the half century from 1940 to 1990, total phosphorus content rose more than 50 percent. This accumulation is a plausible by-product of fertilizer inputs, though trends in "available" phosphorus might have differed from those in total phosphorus content.

The potassium content of Java soils may also have risen since 1940, though the trend is less clear than that for phosphorus. The net rise in K_2O was statistically significant for fallow, not quite significant for cultivated *tegalan* fields, and not at all evident for paddy or tree crop soils. For phosphorus and potassium together, there is both a similarity to China's trends and a difference: In both countries these two total endowments rose overall, but it was potassium that clearly rose in China versus phosphorus in Java.

Soil acidity impedes plant growth in many crops, forcing farmers to choose acid-tolerant crops, such as rice and tree crops, in areas where soil acidity is high. Indonesia has long struggled to adopt liming and water control techniques to promote pH neutrality. On Java, the struggle to raise soil pH into the neutral 6.0–8.0 range has resulted in fluctuating pH values. Table 7.4 and figure 7.3 show that soil pH fell across the World War II decade, then rose from 1950 to 1970. Between 1970 and 1990, however, pH dropped again. The reversal after 1970 resembles the trend switch for organic matter and nitrogen in Java.

This view of the long-run trends in Java's average soil characteristics would not change much if we took into account changes in the proportions of land area devoted to different uses, at least not for the half century after 1940. Between 1940 and 1990 land use changed only very slowly on already settled Java (Booth 1988, chap. 2; Diemont, Smiet, and Nurdin 1991; van der Eng 1996, pp. 280–88). Deforestation did continue in Java, and by itself this would probably have reduced soil organic matter, nitrogen, and potassium. Yet the magnitudes of area changes have not been large since 1940, and the land uses that replaced forest were primarily not the most agricultural or soil depleting. There is little in the trend in land area shares that would alter the picture already sketched for Java in figure 7.3. The same would probably not be true for the earlier period 1880–1940, when crop cultivation and deforestation were still proceeding apace in Java.

Of all the trends just revealed, the most arresting, and therefore the most in need of support from other information, is the impressive buildup of Java's total P_2O_5 and K_2O. Is it really plausible that P_2O_5 should have risen by 44 percent and K_2O should have risen by 28 percent between 1960 and 1990? CSAR soil scientists, comparing samples from the 1980s with earlier survey maps of P_2O_5 and K_2O in the same areas a decade or

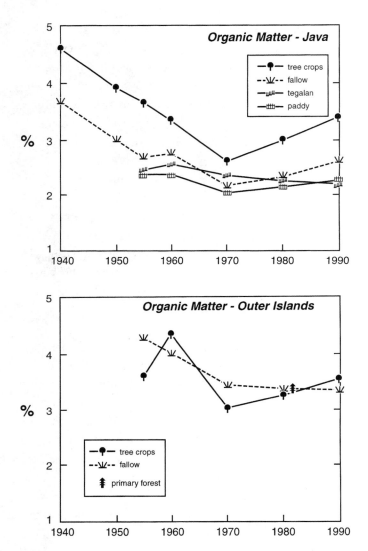

Figure 7.3
Changes in soil characteristics by land use type, Indonesia, 1940–1990

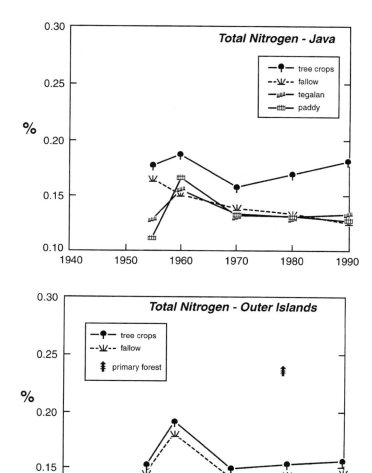

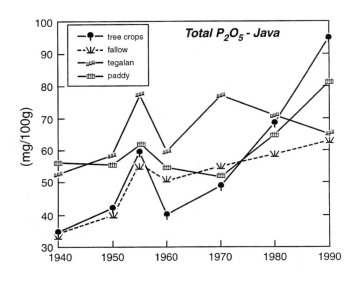

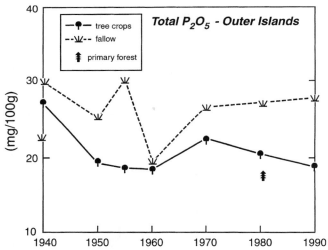

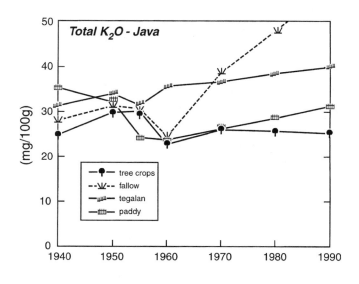

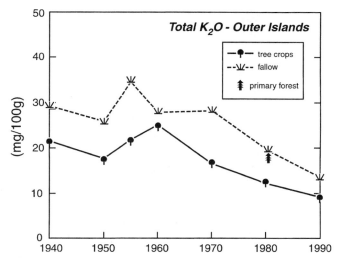

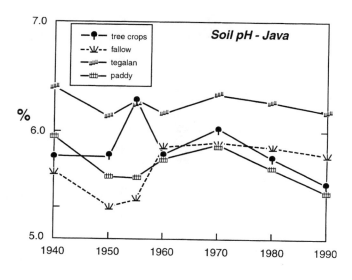

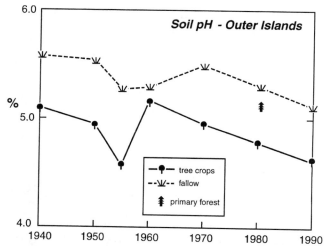

so earlier, have confirmed that P_2O_5 and K_2O levels in the surveyed areas had indeed risen noticeably even within such a limited period (Adiningsih, Santoso, and Sudjadi 1990, pp. 70–72). The applications of fertilizers over this period were indeed substantial for Java, especially for food crops. Given the rapid acceleration of fertilizer applications, it is plausible that Indonesia, like much of China, was able to build up higher and higher phosphorus and potassium content in the soil.

Soil Trends and the Settlement Process on the Outer Islands

The soil trends elsewhere in the archipelago since 1940 stand in contrast to those in Java and Madura. There are many reasons to expect such a contrast, based on the differences in settlement history between the two areas. Java and Madura, along with Bali, have long been densely settled and intensively farmed, whereas the other islands are still frontiers of recent and current settlement. The rate of settlement and cultivation has accelerated since 1970 because of heavy government promotion of migration from crowded Java. This basic settlement motif seems to have played a role in soil trends outside of Java.

Any hints of a net decline in organic matter and nitrogen are more muted for the outer islands than for Java, according to figure 7.3 and table 7.4. One reason is that the paucity of outer-island data for any year before 1955 blocks our view of earlier trends. Since 1955, the data show no statistically significant trends in organic matter and nitrogen for the outer islands, except that fallow lands that were once cultivated showed significantly lower levels of organic matter in 1970–1980 than in 1955. The lack of a clear trend for organic matter over the outer islands as a whole echoes the no-change result that Van Noordwijk et al. (1997) obtained for certain Lampung soils between 1930 and 1990.

Total phosphorus and total potassium did not rise on the outer islands over the period, as they did on Java. On the contrary, potassium has declined for all land uses, at least since 1970. Phosphorus levels in fallow and tree crop soils fell from 1940 to 1960, but showed no trend thereafter.

Soil acidity on the outer islands showed no clear trend up to 1960 or 1970, then increased (i.e., pH fell further below 6.0), holding location

Table 7.4
Changes in soil characteristics by land use type, Indonesia, 1940–1990

Java-Madura	Sample predicted-average soil values					
	1940	1955	1960	1970	1980	1990
Organic matter (%)						
Paddy (*sawah*)		2.36	2.36	2.02	2.12	2.23
Tegalan		2.44	2.57	2.34	2.27	2.19
Tree crops	4.62	3.66[b]	3.34*	2.62**	2.98**	3.38*
Fallow and grasses	3.68	2.67[b]	2.75[a]	2.12**	2.34**	2.58[b]
Weighted average		2.79	2.77	2.34	2.45	2.57
Total N (%)						
Paddy (*sawah*)		0.124	0.156	0.131	0.128	0.127
Tegalan		0.108	0.167**	0.131[a]	0.126[b]	0.122
Tree crops		0.166	0.186	0.155	0.164	0.174
Fallow and grasses		0.157	0.150	0.136	0.128	0.120
Weighted average		0.133	0.169	0.138	0.137	0.137
Total P2O5 (mg/100g)						
Paddy (*sawah*)	55.27	60.93	54.33	50.86	64.19	81.01[a]
Tegalan	52.33	77.51*	59.44	76.81**	70.62*	64.92
Tree crops	34.24	59.68**	40.37	48.79[b]	67.97**	94.68**
Grasses and fallow	32.92	56.30*	50.40*	54.70*	58.17**	61.87[a]
Weighted average	45.31	67.42	52.27	62.40	67.35	75.14

Table 7.4
(continued)

Java-Madura	Sample predicted-average soil values					
	1940	1955	1960	1970	1980	1990
Total K$_2$O (mg/100g)						
Paddy (*sawah*)	35.55	23.83	23.83*	26.19[b]	28.70	31.45
Tegalan	31.31	31.22	35.80	36.81	38.39[b]	40.04
Tree crops	24.93	30.61	22.92	26.30	25.78	25.26
Fallow and grasses	28.17	27.58	24.63	38.81	48.19**	59.82**
Weighted average	29.78	29.49	29.04	32.58	34.71	37.19
Soil pH						
Paddy (*sawah*)	5.95	5.57	5.79	5.89	5.66	5.43
Tegalan	6.40	6.27	6.17	6.35	6.27	6.18
Tree crops	5.77	6.28	5.79	6.02	5.77	5.52
Fallow and grasses	5.61	5.37	5.85	5.90	5.84	5.79
Weighted average	6.06	6.06	5.97	6.14	5.99	5.84

Outer islands	1940	1955	1960	1970	1980	1990
Organic matter (%)						
Tree crops		3.60	4.37	3.01	3.25	3.51
Fallow and grasses		4.32	4.00	3.40*	3.34**	3.29
Primary forest					3.40	

Table 7.4
(continued)

Outer islands	Sample predicted-average soil values					
	1940	1955	1960	1970	1980	1990
Total N (%)						
Tree crops		0.152	0.192	0.148	0.154	0.159
Fallow and grasses		0.145	0.182	0.140	0.141	0.142
Primary forest					0.23	
Total P$_2$O$_5$ (mg/100g)						
Tree crops	27.49	18.58*	18.43*	22.27	20.39[b]	18.77[b]
Fallow and grasses	30.02	30.13	19.28**	26.55	27.09	27.64
Prime forest					17.15	
Total K$_2$O (mg/100g)						
Tree crops	21.45	21.54	24.85	16.59	12.21**	8.99**
Fallow and grasses	29.01	34.85	27.80	27.95	19.24**	13.25**
Primary forest					17.51	
Soil pH						
Tree crops	5.10	4.56**	5.16	4.93	4.77**	4.60**
Fallow and grasses	5.58	5.27*	5.28*	5.49	5.29**	5.09**
Primary forest					5.12	

Sources: Center for Soil and Agroclimate Research soil-profile archive in Bogor, plus a few published studies offering profile data from the 1981–1990 period.

Table 7.4
(continued)

Notes: Cells left blank represent subsamples too thin for reliable estimation of a cell average, even though the soil profiles in many of these cells were usable in the underlying regressions.

Averages shown in each row are conditional means for mixes of places sampled in the 1958–1962 period for the specified land use in either Java or the outer islands. Exception: For primary forest soils, an average for the better-sampled period 1974–1986 was used, with adjustment for the NPK anomaly for 1974–1986 data described in the text.

Each weighted average for Java uses these 1923–1990 sample weights to combine the rows: .149 for paddy, .448 for *tegalan,* .276 for tree crops, and .127 for fallow (shifting cultivation, grasses, and bush).

The (two-tailed) significance of cumulative net changes since the earliest year in the row (usually 1940) is indicated by the following symbols:

**the cumulative change is significant at the 1-percent level;

*the cumulative change is significant at the 5-percent level;

ᵃthe cumulative change is significant at the 6-percent level;

ᵇthe cumulative change is significant at the 7-percent level.

roughly constant. Without controls for location, shifts in the locus of sampling could have greatly distorted the trends, since on any given date acidity was much more pronounced in the wetter north and west than in Nusa Tenggara.

The overall pattern for the outer islands is curious. The period since 1970 has shown some trends that were not present earlier: After 1970 K_2O levels and (for tree crops) P_2O_5 levels, and pH declined, though levels of organic matter and nitrogen did not.

Why should the phosphorus and potassium content have declined in the outer islands since 1970? The question carries possible implications for agricultural performance, since levels of these macronutrients, especially potassium, are low enough in the tropics to constrain yields significantly.

The decline in phosphorus and potassium on the outer islands after 1970 has three possible sources:

1. Cultivating *more marginal lands.* Expanding the cultivated area means sliding down the initial land quality slope.

2. Cultivating *more degraded lands.* Longer cultivation of land might degrade it more and more, in the absence of conservation practices.

3. Cultivating *younger lands.* If longer-worked plots have built up better soil, then a shift toward more recently cleared and cultivated lands could mean a drop in soil nutrients.

The first of these three factors probably played a partial role in the decline of phosphorus and potassium levels on the outer islands. If soil quality, as measured by our soil chemical parameters, were what farmers were trying to maximize initially when choosing a site, then careful initial sorting of sites would mean that expansion of cultivation into sites the initial farmers considered to have poorer soil would cause a drop in inherent soil quality. There are limits, however, to the likelihood that expansion of cultivated areas necessarily brings a loss in quality. In practice other site attributes, such as transportation costs, are likely also to be important in making the choice of initial sites, so that one should not expect that settlers would always choose the sites with the highest soil quality first.

The second factor, arguing that continued cultivation leads to progressively more degraded lands, finds the least support of the three possible sources of phosphorus and potassium decline. Little in the results suggests that there has been a shift toward more soil-degrading practices since 1970.

The third possible factor, shifting to more recently cleared lands, probably accounts for much of the apparent trends since 1970. During that time, the settlement process on the outer islands has accelerated. In effect, the lands sampled on the outer islands in 1990 were apparently *younger,* more frontier lands than were their closest counterparts back in 1970. The trend in outer-island settlement changed around 1974, as shown in figure 7.4. By official measurements, arable lands expanded faster after 1974 than they had expanded before 1970 and faster than the lands that were actually harvested. The change was particularly marked for estate crop lands, meaning those areas deemed suitable for coffee, copra, palm oil, rubber, sugar, tea, and tobacco on large commercial estates.

The shifts in settlement trend made the average site younger in its cultivation by moving two frontiers. First, a greater share of all arable sites were lands that had not been worked or cleared until recently. Second, a greater share of the cleared lands were arable but not yet harvested. It seems very likely that the soil survey sites went through the same transformation as all sites on the outer islands, shifting toward plots with shorter histories of previous cultivation, which shows up even after the soil regressions have controlled for all the measurable physical variables.

The difference that years of previous cultivation could have made can be shown by comparing the outer-island soil averages in figure 7.3 and table 7.4 with those from long-settled Java, for which the physical and source bias variables (the Xs and Ss) would have predicted little net difference in average soil chemistry. As the interisland contrast suggests, more recently cleared sites tend to have soils that are higher in nitrogen and organic matter, lower in total P_2O_5 and K_2O, and more acidic.

It is plausible to advance the joint hypotheses that (a) longer human cultivation may lower organic matter and nitrogen yet raise P_2O_5 and K_2O in tropical soils, and (b) the soils on the outer islands showed the opposite trends after 1970 because the average history of human cultivation of the lands sampled shortened. Figure 7.5 sketches how this might

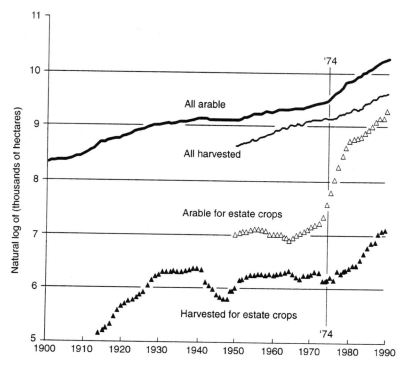

Figure 7.4
Land use trends on the outer islands of Indonesia, 1900–1990

have occurred. For any one humid-tropical site we can envision a life cycle of land clearing and cultivation, as sketched in the upper panel of figure 7.5. After a primary forest site is first cleared, it tends to lose organic matter and nitrogen and, especially in the wettest and warmest areas, potassium. In the first three years after clearing, settlers and tenants typically plant annual field crops, while commercial-crop trees start their growth. Soil fertility drops from its high initial level, and the cultivation of field crops moves on to other sites as the tree crops develop.[8] With time, the now stable tree cultivation enterprise has an incentive to apply fertilizers, and phosphorus and potassium in particular start to accumulate in the soil. As an area's settlements got older, we would imagine that soil nutrients would build up.

On the outer islands since 1970, however, the average age of the settlements got younger, as sketched in the lower panel of figure 7.5. A

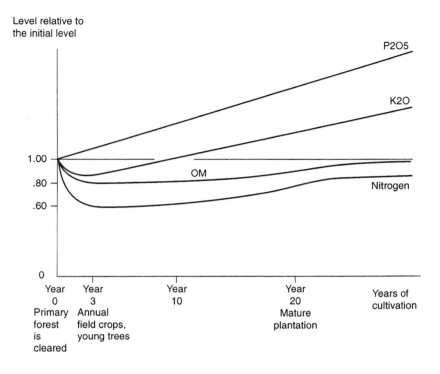

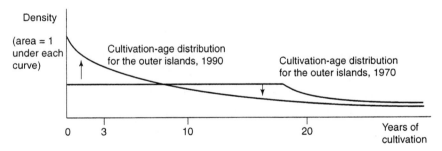

Figure 7.5
Schematic sketch of possible soil chemistry dynamics in a tropical primary forest cleared for commercial tree crops

greater and greater share of lands were in the early nutrient-depleted phase that comes right after clearing and the start of cultivation. This is likely to have been a main reason why phosphorus and potassium in particular declined so much after 1970.[9]

Signs of Erosion?

The main fear about soil degradation in Indonesia, as chapter 1 noted, has been that human practices have increased the rate of soil erosion. The historical data allow two kinds of indirect tests of the broad overall effects of erosion on Indonesia's topsoils, one chemical, the other physical.

First, one can make use of the widely noted effects of typical erosion on topsoil endowments of all the macronutrients and of organic matter (Follett and Stewart 1985, pp. 339–41; Larson, Pierce, and Dowdy 1983; El-Swaify, Moldenhauer, and Lo 1985, chaps. 22, 23, 28, 29). If erosion has been the dominant form of topsoil degradation, one would expect to see a downward trend in total nitrogen, phosphorus, potassium, and organic matter, as noted in chapter 1. Figure 7.3 and table 7.4 give no such result for Indonesia's cultivated areas, however. The most obvious contradiction they offer to any central role for erosion is the rise in topsoil phosphorus and potassium in intensively cultivated Java. The timing of the decline in organic matter and (probably) nitrogen also works against the conventional emphasis on erosion. If accelerating erosion is the main source of soil degradation in Indonesia, why did the declines in organic matter and nitrogen occur before 1970, with no further decline after 1970, during the intensification of Java's agriculture and the settlement boom on the outer islands?

A second, physical clue to erosion is the thickness of the topsoil layer. Some soil scientists have suggested that losses from erosion should show up over time as topsoil thinning. This possibility can be tested for Indonesia, in the same way that chapter 4 tested it for China.

Table 7.5 offers crude topsoil thickness averages for different data periods. At face value, the thickness estimates suggest topsoil losses between c. 1940 and c. 1990 in Java, on fields that the data explicitly describe as cultivated. But the top-horizon thicknesses should not be

Table 7.5
Ostensible changes in thickness of topsoil layer in Indonesia, 1923–1990

| | Average thickness of top horizon | | Share of profiles with bottom depth (in cm) ending in 0 or 5 |
	All sites	Cultivated sites	
Java			
1923–1935	24.1	na	96%
1936–1944	19.6	19.8	94%
1947–1957	22.0	23.0	84%
1958–1967	18.1	18.0	77%
1968–1973	18.7	18.8	64%
1974–1986	17.6	17.8	41%
1987–1990	17.6	17.7	41%
Outer islands			
1926–1937	15.7	na	66%
1938–1942	17.1	16.9	75%
1947–1957	17.4	18.9	76%
1958–1967	16.1	17.5	72%
1968–1973	17.1	17.2	62%
1974–1986	13.6	13.6	39%
1987–1990	16.6	16.1	38%

Note: "Cultivated" includes *sawah*, *tegalan* field crops, and tree crops. It excludes all pasture, grasslands, and primary forest.

taken at face value. The earlier data do not represent conventional soil horizons. Rather, they reflect horizons down to arbitrary depths, and the purpose of sampling may have become less geological (deep readings) and more agronomic (shallower readings) over time. A clue to the arbitrariness of the horizon depths is the high proportion of topsoil bottom depths that are rounded off to the nearest ten or five centimeters, as table 7.5 also shows. Statistical regressions (not shown here) confirm that there is less to the thinning of Indonesia's topsoils than meets the eye. When one includes an independent variable that equals one for all profiles having a zero or five as the last digit on the depth of the bottom of the topsoil horizon, that variable is a highly significant contributor to apparent thickness. Controlling for it, there are no significant trends in soil thickness since the 1920s, either in Java or on the other islands. What might look like a net loss of topsoil may therefore just be a shift toward more precise definitions of soil horizons in terms of characteristics. This suggestion is the same as the one made for China in chapter 4.

The Geography of Indonesia's Soil Chemistry

A 1990 Snapshot

The Indonesian soil database can shed light on how soil conditions vary over space as well as over time. By adjusting all the soil characteristics from the sampling data up to a common year, 1990, table 7.6 sketches an overall soil geography for Indonesia, revealing some broad relationships of soil characteristics to climate and to human history across the archipelago.

Table 7.6 and figure 7.6 show some broad basic patterns in that geography. One influence that shapes the soil differences between islands is the number of months of rainfall. As we have seen earlier, precipitation in Indonesia is heavier to the west. As soil scientists would predict, in a tropical setting the extra moisture brings heavier vegetation, greater acidity (lower pH), and a quicker loss of potassium. Table 7.6 reveals patterns that fit these expectations, if one examines the data for the islands and provinces more or less longitudinally, from the west (top rows) to the east (bottom). The K_2O content and pH are indeed lower in the wetter west. On the other hand, organic matter levels run higher in the west, where they are even higher in relation to nitrogen. These gradients from west to east, though unmistakable, are not smooth and monotonic, however. Population density exerts a separate influence that makes Java noticeably richer in P_2O_5 and K_2O than other islands to either the east or the west. Irian Jaya also has a moist climate, despite being at the east end of the country.

Particularly systematic in its geography and important for its potential implications for crop choice is the pattern of pH and acidity[10] shown in figure 7.6. Again, this pattern tends to follow the rate of precipitation across the islands. Acidity is of concern for agriculture primarily because it limits the choice of crops that can be grown in a particular area. Rice and most tree crops are quite acid-tolerant, and highly acid soils tend to push farmers toward these crop choices. Should relative prices for rice and tree crops take a downturn, there would be pressure to alleviate the soil acidity, perhaps with liming techniques developed to suit Indonesian conditions. Acidity is reversible, so a broad range of vegetables can be grown on lands where acidity would otherwise check their growth. But there is a cost.

Were Java's Soils Always Different?

The magnitude of the differences in the soil trends between Java and the outer islands invites the conclusion that they are the results of human soil amendments alone. The buildup of total P_2O_5 and total K_2O in particular has seemed to reflect the accelerating application of fertilizers. The high level of these nutrient reserves seems to support the optimistic generalization recently voiced by Alan Strout (1983) on the basis of his study of Java soils: "It should always be recalled that differences in soil fertility and climate can usually be more than compensated for by man's own intervention" (p. 49). To the extent of finding compensating ways to raise yields, this is surely true. The world abounds with cases in which yields have been raised dramatically in the face of poor soils and poor climate. Often the achievement comes at a high price in resources, as demonstrated by Saudi Arabia's costly achievement of exportable surplus wheat in the 1980s. Yet the trends in Java's soils extend Strout's dictum even further by suggesting that humans' pursuit of higher yields through intensive agriculture also raises the nutrient reserves in the soil. Is that what happened on Java? Could a soil scientist rightly classify Java's richer soils as "anthropogenic"?

The opposing idea here is the suspicion that Java's soils were richer from the start in some key nutrients, prior to any human interventions. Perhaps the causation works the other way: People have crowded into Java, and grow more food per hectare there, because Java's soils were always better. Perhaps the richer nutrients in Java's soil cause the higher agricultural intensity there, not vice versa. Strout (1983) has also offered this opposing conjecture: "Fertiliser use is concentrated on Java and the province of North Sumatra, areas where inherent soil fertility is believed to be high. . . . [I]t would appear that natural soil and climate factors on Java and Bali may be roughly three times as favourable as [those] found in much of Kalimantan and Maluku" (p. 48). Can one have it both ways?

Table 7.6
Geography of soil chemistry in Indonesia: Estimated provincial and regional averages, 1990

	Organic matter (%)	N (%)	P$_2$O$_5$ (mg/100g)	K$_2$O (mg/100g)	pH	Acidity
Aceh	3.28	0.156	21.71	23.17	4.78	1.22
North Sumatra	5.09	0.236	25.91	23.20	4.88	1.12
West Sumatra	5.13	0.243	27.81	16.12	4.99	1.01
Riau	3.92	0.171	14.00	9.09	4.52	1.48
Jambi	3.18	0.140	11.86	10.34	4.15	1.85
South Sumatra	5.17	0.222	19.28	12.25	4.78	1.22
Bengkulu	4.44	0.235	27.45	17.94	4.86	1.14
Lampung	4.27	0.213	23.98	11.34	5.08	0.92
West Kalimantan	5.01	0.218	14.71	8.58	4.45	1.55
Central Kalimantan	5.17	0.236	25.52	9.66	4.39	1.61
South Kalimantan	5.14	0.285	29.42	9.51	4.69	1.31
East Kalimantan	4.53	0.216	23.53	12.60	4.78	1.22
West Java (including Jakarta)	3.43	0.159	80.34	36.90	5.38	0.68
Central Java (including Yogya.)	2.56	0.119	90.19	64.92	6.15	0.23
East Java (including Madura)	2.39	0.117	68.75	63.69	6.33	0.09
Bali	7.19	n.a.	219.21	148.45	5.84	0.25
West Nusa Tenggara	2.61	0.127	116.43	85.31	6.44	0.02
East Nusa Tenggara	1.80	0.099	100.36	113.89	7.20	0.00
North Sulawesi	2.81	0.157	29.11	41.11	5.48	0.54
Central Sulawesi	2.93	0.177	25.82	23.41	5.99	0.13

Table 7.6
(continued)

	Organic matter (%)	N (%)	P₂O₅ (mg/100g)	K₂O (mg/100g)	pH	Acidity
South Sulawesi	2.44	0.181	74.52	44.02	5.18	0.83
Southeast Sulawesi	3.06	0.185	30.87	15.66	5.44	0.56
Maluku	6.36	0.331	49.98	24.75	5.89	0.19
Irian Jaya	2.61	n.a.	51.57	28.07	6.08	0.07
Totals:						
Sumatra	4.57	0.209	20.54	13.91	4.77	1.23
Kalimantan	4.91	0.227	20.02	10.05	4.56	1.44
Java, Bali, Madura	2.91	0.135	82.22	54.58	5.89	0.37
Sulawesi	2.78	0.179	47.73	31.09	5.50	0.54
East (Nusa Tenggara, Maluku, Irian Jaya)	3.07	0.166	82.57	65.14	6.42	0.07

Notes: For organic matter, N, P₂O₅, and K₂O, averages are slightly lower than conventional averages would be for a procedural reason. At the district (*kabupaten*) level, averages are geometric averages, not conventional arithmetic averages. That is, they are antilogs of weighted sums of logs, not weighted sums of the original measures. At the district level, they will therefore run lower than the conventional ones, giving more implicit weight to cases where organic matter or N or P₂O₅ or K₂O are low. From the district level up to the provincial level or higher, however, the averaging is conventional.

Note that these averages depart from those in figure 7.3 and table 7.4, which are averages of predicted values by land use type, with sample weights from 1958–1962.

Figure 7.6
Average predicted pH by district, Indonesia, 1990

- = 650 or higher
- = 6.00 – 6.49
- = 5.50 – 5.99
- = 5.00 – 5.49
- = 4.5 – 4.99
- = under 4.50
- na = sufficient data not available

Can it be true that Java's inherent soil advantage causes the higher intensity of its agriculture, which in turn is responsible for its higher yields and even its extra soil endowments?

In fact, it does seem that one can have it both ways, by splitting the difference between the two views. Java's soil history does testify in part to the power of humans to augment soil endowments, particularly the total contents of phosphorus and potassium. That is already evident in those strong up-slopes for P_2O_5 and K_2O on Java (figure 7.3). At the same time, although humans added nutrient reserves to Java's soils since 1940, Java's soils were not the same as those of other islands in 1940 or earlier. Two kinds of clues imply that Java has always had soils richer in phosphorus and potassium than the other islands of Indonesia. One is the fact that the P_2O_5 and K_2O endowments were already clearly higher on Java than on the outer islands in 1940 (again, see table 7.3). By itself, this first clue need not be accepted as strong evidence against the human origins of Java's soil nutrients, since Java was already settled and cropped well before 1940. It receives further support, however, from a second clue, one relating to forest soils. Soils sampled from sites labeled "primary forest" (*hutan primer*) seem to be virgin soils with no history of cultivation. Although there are only a few such sites on Java, they do lend themselves to a comparison with some of the more abundant primary-forest soils on the outer islands. The averages are shown in table 7.7. To the extent that primary-forest sites received no human inputs, the comparison in table 7.7 hints at the prehuman soil chemistry of the different islands. The contrast between the forest soils in the two areas suggests that Java indeed had better soils from the start. Java's P_2O_5, K_2O, and even its organic matter were higher than in the outer islands, and acidity was less pronounced in Java's forests. To some degree, then, Java had soil advantages from the start, presumably thanks to geology and climate.[11] Yet as we have seen, the interisland gap in P_2O_5 and K_2O widened greatly after 1940.

Summary

Thanks to a data set developed in the early 1990s with the cooperation of Indonesia's CSAR, it has become possible to trace the trends in

Table 7.7
Average levels of nutrients in forest soils, Indonesia

	Forest soils on Java, all dates	Forest soils on the outer islands, 1974–1986 (from table 7.4, $n = 133$)
Organic matter (%)	4.58 ($n = 35$)	3.40
Nitrogen (%)	0.22 ($n = 25$)	0.23
P_2O_5 (mg/100g)	33.6 ($n = 33$)	17.2
K_2O (mg/100g)	44.9 ($n = 36$)	17.5
pH	6.14 ($n = 20$)	5.08
Acidity	0.37 ($n = 20$)	1.09

Indonesia's topsoil chemistry from 1940 to 1990. Considerable caution is required in working with data from so many different places and times, but the data suffice to judge some of the broad arguments about what has been happening to the soil in Indonesia in the last half century.

In Java, a decline in some soil endowments but not in others has accompanied the rising intensification of agriculture. Organic matter declined on cultivated Java soils from 1940 to 1970, then rose slightly by 1990. Total phosphorus and potassium have risen over the half century as a whole. Acidity showed cycles, with no clear overall change on Java. Whether this represents a net gain or loss in soil quality depends on the weights one attaches to different soil inputs.

On the outer islands phosphorus, potassium, and pH all decreased, especially after 1970. Trends in organic matter and nitrogen were less distinct. The fall in phosphorus and potassium levels may partly reflect a shift in the outer islands to younger, less cultivated sites due to the acceleration of human settlement and cultivation from about 1974 on.

The contrast between Java and the other islands at any recent point in time offers a significant clue about human impact on soil chemistry. Java is agriculturally old and crowded, whereas the average site on the outer islands has stayed young. In any given year, soil characteristics differ sharply between Java and the outer islands in a way that seems consistent with the human imprint visible in trends both in Indonesia and in China. Longer settlement and denser cultivation may deplete organic matter and nitrogen, yet they raise total phosphorus and potassium and reduce acidity.

Analysis of the available data failed to show that erosion was a key source, or an accelerating source, of soil degradation in Indonesia over this half century. If erosion were a key source of degradation, we would expect it to have reduced total phosphorus and potassium, but that seems not to have happened. Additionally, the decline in organic matter and nitrogen was over by 1970, before the events that should have accelerated the rate of soil loss due to erosion, namely, the intensification of cultivation on Java and the expansion of forest clearing and settlement on the outer islands. Furthermore, what looks at first like physical evidence that topsoils have gotten thinner might have reflected simply a change in soil survey purpose and practice, not a true thinning of the topsoil. Perhaps research on soil degradation in Indonesia should concentrate less on erosion and its effects and more on other human-induced processes, such as fertilizer application, water control, and nutrient depletion through crop uptake.

8

Consequences for Indonesian Agriculture

What happened to Indonesia's soils across this century must have affected the nation's ability to supply itself with food and other agricultural products. Various soil changes revealed by chapter 7, each significant in itself at times, must have pulled in opposite directions. Between 1940 and 1970 on Java, for example, the apparent decline in organic matter and possibly in nitrogen should have lowered production, whereas the rise in phosphorus and potassium should have raised it. How can we weigh these opposing trends against each other to determine their net impact on agricultural production? That is, has Indonesian soil quality, the agricultural potential of its soils available to agriculture, been improving or deteriorating over the period studied?

As long as soil quality matters to agriculture, but not vice versa, the road to an answer is clear: We need to add soil conditions to the list of variables that determine agricultural production and use statistical analysis to quantify the effect of soil conditions, holding other agricultural inputs equal. This could be done on experimental plots, for example, by making sure that only the soil characteristics varied from plot to plot. Better yet, we can look at actual experience on actual farm plots, using statistics to determine how the effects of many things sort themselves out simultaneously in the real world.

This chapter explores the links between soil and agriculture in a cross section of places in 1990, both for Java and for the rest of the archipelago. We compare places instead of comparing times for a very practical reason: There are more districts (*kabupaten*) to compare than there are years of annual data consistently defined. The spatial comparisons will yield a host of results, the key ones being the impacts of soil chemistry on

Approach A. All Agricultural Inputs are Exogenous.

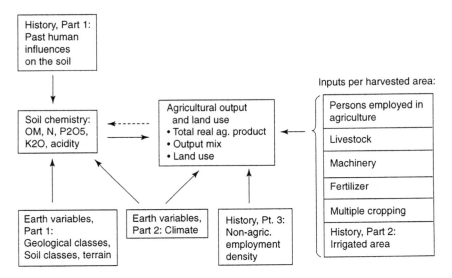

Figure 8.1
Causal system linking soil chemistry and agricultural productions in Indonesia

yields. Armed with these impact measures, this chapter will then be able to measure the economic value of the soil changes revealed in chapter 7 by posing the following question: Given the marginal effects of soil quality on agriculture and vice versa in 1990, how much economic difference would it have made if Indonesia in 1990 had faced its soil endowment of, say, 1940 instead of 1990?

Alternative Approaches

The road to sorting out the contribution of soils to agricultural productivity must follow a different route in Indonesia from that taken in chapter 5 to estimate the simultaneous soil-agriculture system in China. To understand why, let us contrast two approaches to describing and estimating the soil-agriculture system, one that tries to adapt chapter 5's China system to Indonesian data, and one that reflects the need for a fresh departure.

The two panels of figure 8.1 sketch the two approaches. The arrows represent likely causal influences that need to be quantified. The two sets of variables to which the arrows point—namely, soil chemistry and

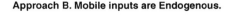

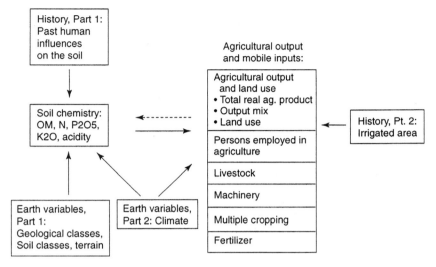

Figure 8.1 (continued)

agricultural outputs—are dependent variables at the center of a system that must be estimated statistically.

Approach A in figure 8.1 closely resembles chapter 5's picture of the soil-agriculture system in China. It is an approach that occurs naturally to the agricultural economist, for whom a first step proposed by approach A is to incorporate some soil chemistry characteristics into the well-known list of inputs in the agricultural production function. On this view, soil is one of many inputs into production. The other inputs, displayed on the right, are more easily shaped and accumulated by humans than is soil. What chapter 5's model added for China, and approach A tries to add for Indonesia, is the reverse (dashed) arrow running left from agricultural outputs back to the soil. Knowing that agriculture and soil affect one another, the economist next tries to estimate the whole system by making use of the fact that some things directly affect soil but not agriculture, and others directly affect agriculture but not soil. Knowing that the causal forces on the left side of the first panel of figure 8.1 affect soil, whereas the inputs on the right affect agricultural output and land use, we can identify all the separate relationships statistically. Every arrow can be quantified. Chapter 5 did that for China.

Indonesia, however, poses a problem that did not arise for China in the 1980s concerning the list of "inputs per harvested area" on the right-hand side of the figure. These are not independent forces that vary across the districts of Indonesia the way they do across the counties of China. For China, chapter 5 could legitimately argue that government decisions at the national and provincial level virtually dictated the separate supply of key inputs to each county (*xian*) up through the mid-1980s. Tight restrictions on human migration and the rationing of fertilizer and capital meant that the county's supplies of these inputs could be taken as exogenous independent variables in determining agricultural output. As long as they are exogenous, these input variables on the right side of approach A in figure 8.1 help us quantify the whole system, including the causal arrows in both directions between soil chemistry and agricultural outcomes.

Identifying the whole simultaneous system cannot work the same way if inputs are mobile between the places in the sample. Consider the extreme case in which laborers, capital, fertilizer, and all other nonland inputs can move freely between any two places in response even to small differences in their rewards. Suppose, for an extreme example, that an agricultural worker in East Java can migrate to North Sumatra or South Sulawesi or distant Biak whenever the pay looks better there. In such an integrated national labor market, there *is no* local labor supply, there is only the same supply of the whole nation's potential workers to agriculture in every district. The same could be true of fertilizer or capital, which could be available to every district in the country at nearly uniform prices. If there is a single national market, then it is misleading to imagine that local input supplies determine local agricultural outputs. Rather, profitable opportunities simultaneously determine where the inputs go and how much they produce. Local outputs and inputs are determined together, instead of the local input supplies' causing local outputs. Such integrated factor markets are a partial reality in Indonesia, aided by government subsidies to outward migration from Java and by subsidized government fertilizer supplies and loans.

Once most of the inputs into agriculture become mobile between places—even just partially mobile—we can no longer use approach A to repeat the simultaneous-equation analysis of chapter 5. If inputs are mobile, aside from land and irrigation infrastructure, a cross-sectional contrast of places in the same year fails to reveal how agriculture affects soil (the dashed arrow). Agricultural labor and other inputs are no longer independent variables, but basically just a response to soil, climate, other physical forces, and random local influences. The dashed line cannot be drawn on the basis of a national cross section of places.

To quantify the soil-agriculture nexus in a nation or region with integrated markets, one must remove from the cross-sectional analysis the dashed arrow from agriculture to soil and find some separate way to sort out how humans have affected the soil. In all that follows, we must break the task into two separate parts:

1. Estimate how human *agriculture affects soil* by using historical time as the proxy for all past human impact on the soil. This was already done in chapter 7.

2. Estimate how *soil affects agriculture* by using a reduced-form single-equation approach in which there is no simultaneous feedback from local agriculture to local soils to complicate the measurement of soil's effects on agricultural output and land use. The current chapter undertakes this task, using a cross section of Indonesian districts in 1990.

To take the second step correctly requires using measures of soil chemistry that are free from any same-year feedback from agricultural practice to the condition of the topsoil. We do this by taking measures of soil condition that are *predicted values* from chapter 7's regression equations, in which the influences of figure 8.1's "Earth variables" and "History, Part 1" dictate our measures of topsoil chemistry. Such measures are not biased by any feedback from current (here, 1990) agricultural practice, both because 1990 data accounted for only a tiny share of chapter 7's historical soil samples and because all local randomness has been purged from the predicted values of soil organic matter, N, P_2O_5, K_2O, and acidity.

Market integration therefore alters how we portray and quantify the soil and agriculture system. Consider a useful alternative portrayal, approach B, in the second panel of figure 8.1. This portrayal recognizes that mobile agricultural inputs are determined simultaneously with agricultural output and land use. That leaves almost no local inputs predetermined by policy or past history. On the right-hand side of the figure, we have only the legacy of built-up irrigation infrastructure. To this one could have added the local presence of nonagricultural employment opportunities, but even this should probably be excluded from the list of exogenous inputs if laborers are highly mobile between agricultural and nonagricultural jobs.

Which promises more accuracy, approach A's assumption that all local markets are separate, or approach B's assumption that all of the districts in a sample share the same integrated market? The truth lies somewhere between the two. To the extent that factor markets are not perfectly integrated in Indonesia, there might still be some basis for approach A's imagining that each district has its own labor supply, its own fertilizer supply, and its own capital supplies. But overlooking factor mobility might still lead to biased results. It is useful to show the results of both approaches side by side, and to reach conclusions that blend the two.

Even when we use approach A, however, it is not wise to try to imagine that Indonesia's spatial patterns in 1990 can reveal how agricultural outputs and inputs affect the soil, as was possible for China in chapter 5. A wiser choice is to abandon simultaneous-equation methods here, letting chapter 7's soil history determine how humans have affected the soil over time. There is a clear logic to using a different procedure for the two countries, based on the realities of factor markets. Factors of production cannot migrate over time in response to time differences in their rewards. To imitate how history makes the separate economic conditions of different decades affect the condition of the soil, it is better to find a spatial laboratory in which factors cannot move, and therefore conditions are separate, from place to place. China was such a place in the mid-1980s, but Indonesia is not. For Indonesia, the agriculture → soil analysis must be historical, as mentioned, whereas a spatial analysis can trace the soil → agricultural relationship here.[1]

The Impact of Soils on Agricultural Productivity in 1990

The approaches just introduced illuminate the role of soil chemistry and other factors in determining the productivity of Indonesian agriculture, which displays a geography both of uneven productivity and of uneven prosperity as well. Figures 8.2 and 8.3 show the spatial inequalities in productivity and prosperity that need to be explained by differences in soil, climate, and other factors. Figure 8.2 shows where agricultural output per harvested area, or land productivity, was highest (the top, or darkest, 20 percent of districts) and lowest (the bottom, or white, 20 percent of districts) in 1990. In general, output per harvested area was greatest where population is most dense, particularly in Java and Bali. The land productivity patterns in figure 8.2 might, if viewed alone, seem to make only the familiar point that adding extra agricultural laborers to the same land raises yields.

Yet the geography of output per agricultural laborer, or labor productivity, contradicts any simple notion that low labor productivity plagues the agriculture of those crowded areas that had high land productivity. On the contrary, figure 8.3's view of labor productivity in 1990 is far from the opposite of the pattern of land productivity. In fact, the areas with the highest agricultural labor productivity are much the same as those with the highest land productivity. Roughly speaking, these areas are simply more prosperous than those that tend to have low productivity by both measures. Topping the labor productivity ranks are again Bali and Java, particularly West Java, joined by South Sulawesi and the area of North Sumatra near Medan. We need to explain, then, something more interesting than mere differential crowding of different islands. Why are some areas much more productive and prosperous than others by either crude measure? Is it all a story of agriculture's being more productive where it is near the centers of the non-agricultural economy? Or do differences in soil quality play a role?

The testing ground used here to explore these questions is the cross section of Indonesia's districts (*kabupaten*) in the year 1990. The sample includes all districts for which we have soil profiles in chapter 7's historical sample, allowing use of regression-predicted average soil characteristics:[2] 77 of the 82 districts of Java, and 96 of the 147 districts of the outer islands, for a total of 173 sample districts.

Both the all-Indonesia sample of 173 districts and the separate samples for Java and the outer islands have their uses. The all-Indonesia sample shows the net results of three kinds of geographical variation: variations among the districts of Java, variations among the outer-island districts, and differences between Java as a whole and the outer islands as a whole. Although our main focus should be on the net effect of all three

Figure 8.2
Gross value of agricultural production per harvested area, 1990

Irian
Jaya

Maluku

N. Sulawesi

C. Sulawesi

E.
Kalimantan

S.E.
Sulawesi

East
Nusa
Tenggara

East Timor

West Nusa
Tenggara

S. Sulawesi

W.

S. Sumatra

Lampung

Riau

Bengkulu

North
Sumatra

West
Sumatra

Aceh

■ = top quintile (over 1600)

▨ = 2nd quintile (1275 – 1599)

▦ = middle quintile (1000 – 1274)

▥ = 4th quintile (750 – 999)

□ = bottom quintile (under 750)

(1000s of Rupiahs per harvested hectare)

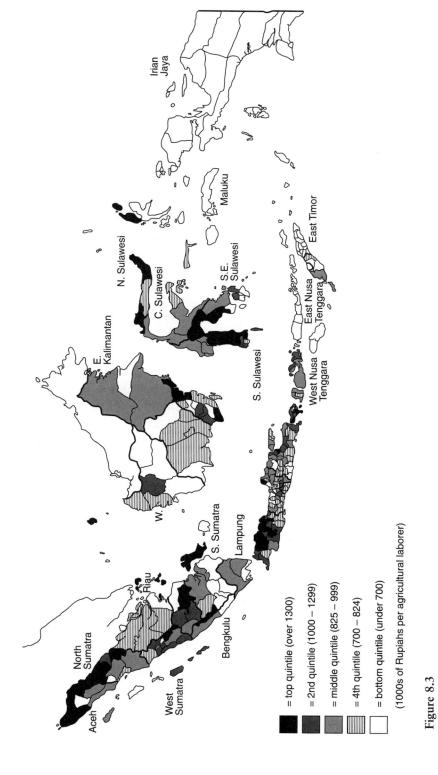

Figure 8.3
Gross value of agricultural production per laborer, 1990

variations, and therefore on the all-Indonesia results, the separate patterns observed inside and outside Java can help decompose the overall results into geographical parts.

The production function used to explain differences in Indonesia's agricultural productivity is of a simpler form than that used for China in chapter 5, even with approach A's fuller list of agricultural inputs.[3] True to the sketch in figure 8.1, approach A's fuller equations relate output per harvested area in 1990 to the inputs of agricultural employment, work animals, machinery, fertilizer, and arable land per hectare harvested, plus water supply and soil chemistry. Where possible these are expressed in logarithms, as in the Cobb Douglas production function.[4]

The agricultural output and input variables are measured per unit of land area, as in chapter 5 and in many other studies of agricultural productivity. The choice of a land area denominator is different for Indonesia, however, than in chapter 5 or other studies. Most countries, like China, measure sown or harvested area and cultivated area, the latter being all land area that entered production at least once during the year. The ratio of these—(sown or harvested area/cultivated area)—serves as the MCI, a familiar measure of the intensity of cultivation that was also used in chapter 5. One of the two widely available measures in Indonesia's agricultural statistics is harvested area, a conventional definition. Indonesia's other available measure, arable land, has a more elusive meaning, however. Much of what counts as arable, especially on the outer islands, is land not actually in production, though it is sufficiently cleared that it could be brought into productive use without much difficulty. If we were to measure productivity as the value of output per hectare of arable land, we would be artificially lowering the measure of district productivity. That would be harmless if the ratio of arable to harvested land varied only for the right reason, namely, in being lower in years when there was more multiple cropping. But output per arable area can be raised or lowered misleadingly, just by counting less or more idle land as arable.

For ease of interpretation, therefore, let us measure output and the inputs of labor, work animals, machinery, and fertilizer per unit of *harvested* area here. The fuller equations of approach A also use arable area per harvested area to pick up three effects at once: the multiple-cropping rate (harvested area/unknown true cultivated area), the unknown partial use of extra arable land in production, and any possible selectivity effect whereby greater availability of arable land puts a better selection of local soils into production.

Table 8.1 shows the agricultural productivity patterns revealed by the two approaches. Let us focus first on approach A's fuller and more conventional production function analysis, starting with the input measures at the top. The top row of the table gives a result that is easy to understand, though it may surprise us at first. At face value it seems to say that agricultural employment has no effect on output per land area, as if Indonesia had a labor surplus, with zero marginal product for labor in the countryside. Recall, however, that we are measuring outputs and inputs per hectare of harvested area. Therefore the near-zero labor coefficient throughout Indonesia suggests that the marginal product of labor is fully reflected in its effect on the harvested area, and not through more intense working and more frequent harvesting of each hectare cultivated.

The next three inputs in the table—work animals, machinery, and fertilizer—show positive productivity even beyond their effects on the number of harvests per year, both for the whole Indonesian archipelago and among the islands outside of Java. Although this pattern is highly significant overall, note that it fails to hold for Java. Perhaps the near-zero effects for Java show that in Java's intensive agriculture, the productivity of work animals, machinery, and fertilizer, like the productivity of extra laborers, is fully manifested in the ability to harvest more often each year. It also seems likely that a mixture of its greater homogeneity and some measurement errors keeps the Java coefficients closer to zero.[5] The districts of Java show less variation than the outer-island districts in their agricultural outputs and inputs per harvested area. For Java it may be harder to show the positive effects of different inputs, especially if any unknown measurement errors bias coefficients toward zero.

The effect of extra arable land, per hectare harvested, is strongly negative on the outer islands and for Indonesia as a whole, though it is somewhat positive for Java. Historical patterns in the settlement process and the statistical reporting from the outer islands may explain this. For the outer islands, the fact that settlement is more recent and continuing seems to explain not only why so much land is reported as arable but not heavily harvested, but also why the yield per harvested hectare may be

Table 8.1
Regression equations for the determinants of agricultural output per harvested area in Indonesia's districts in 1990

Dependent variable = [ln (million Rupiahs of agricultural output per harvested hectare)]

Approach A. All agricultural inputs treated as exogenous

Independent variables	(1) All Indonesia		(2) Java		(3) Outer islands	
	Coeff.	|t|	Coeff.	|t|	Coeff.	|t|
Inputs per harvested area						
ln (agricultural employment)	−0.0761	(0.98)	0.0480	(0.43)	0.0619	(0.61)
ln (work animals)	0.0868	(3.62)**	−0.0607	(1.43)	0.0939	(3.16)**
Machinery value	0.0054	(1.57)	0.0012	(0.37)	0.0136	(2.52)*
Fertilizer (nitrogen content)	0.0013	(2.10)*	0.0009	(1.56)	0.0054	(4.29)**
ln (arable area)	−0.1198	(2.13)*	0.2542	(1.59)	−0.1540	(2.37)*
Environmental influences						
Water	0.4362	(1.63)[b]	−0.1109	(0.19)	0.8591	(2.55)*
Water squared	−0.1936	(0.57)	−0.2398	(0.34)	−0.8949	(2.18)*
Nonagricultural employment	0.1764	(4.49)**	0.2209	(4.30)**	0.0332	(0.51)
Soil chemistry						
ln (organic matter)	0.1175	(1.64)[b]	0.1255	(1.51)	0.0415	(0.42)
ln (phosphorus)	−0.0041	(0.08)	−0.0252	(0.33)	−0.1055	(1.45)
ln (potassium)	0.1253	(2.01)*	−0.0107	(0.12)	0.1526	(1.95)[a]
Acidity	0.0794	(1.05)	−0.0014	(0.01)	−0.0602	(0.66)
Constant	−0.5363	(2.27)*	0.3016	(0.98)	−0.2626	(0.82)

Table 8.1
(continued)

Adjusted R-squared	.547	.460	.576
Dependent variable mean	0.1294	0.3684	-0.0624
Standard error of estimate	0.2812	0.1929	0.2741
Number of districts	173	77	96

Approach B. Mobile inputs treated as endogenous

Independent variables	(4) All Indonesia Coeff.	|t|	(5) Java	(6) Outer Islands Coeff.	|t|
Soil chemistry					
ln (organic matter)	-0.0616	(0.80)	[*Results of Equation (5)*	-0.100	(0.79)
ln (phosphorus)	0.1474	(2.69)**	*were*	0.164	(1.91)a
ln (potassium)	0.1226	(1.75)b	*insignificant.*]	0.147	(1.43)
Acidity	-0.0916	(1.14)		-0.074	(0.63)
Other influences					
Water	1.2974	(4.83)**		1.366	(3.14)**
Water squared	-0.9246	(2.51)*		-1.017	(1.89)a
Constant	-1.0121	(4.16)**		-1.121	(2.93)**
Adjusted R-squared	.394			.237	
Dependent variable mean	0.1294			-0.0624	
Standard error of estimate	0.3251			0.3678	
Number of districts	173			96	

Table 8.1
(continued)

Sources: All agricultural output and input data were collected by Pierre van der Eng from the Indonesian government's annual volumes on land use, production, fertilizer sales, and labor, plus the agricultural census.

Notes: Statistical significance of coefficients:

**1-percent level; *5-percent level; [a]6-percent level; [b]7-percent level.

Fertilizer inputs are reported only according to their nitrogen content. Separate figures were usually available for phosphorus and potassium contents as well, but these were highly correlated with the nitrogen content when they were reported. To avoid multicollinearity, these equations use the nitrogen content to represent all three nutrients (nitrogen, phosphotus, potassium) in fixed proportions.

Units of measurement for independent variables: Agricultural laborers—persons employed primarily in agriculture, per harvested hectare; work animals—work animals per harvested hectare; machinery value—millions of Rupiah per hectare; fertilizer (nitrogen content)—millions of Rupiah per hectare; arable area per harvested area—hectares per hectare; organic matter—percentage; phosphorus—mg of total P_2O_5 per 100 g of soil; potassium—mg of total K_2O per 100 g of soil; acidity—units of pH below 6, averaged over sampled soils with acidity—min $(0, 6 - pH)$ for each individual sampled site; water—effective net number of wet months or availability of irrigation (see below); nonagriculture employment—persons employed outside of agriculture, per harvested hectare. Unlike the other inputs, machinery and fertilizer could not be given in logarithmic form because zero values were recorded for these two inputs in a few districts on the outer islands. Water supply variable—(Irrig.) + [(1 − Irrig.) * (net wet months)/12], where Irrig. is the share of all harvested area that was irrigated. The net wet months variable is based on a 1:250,000 map of precipitation patterns, generously supplied by the Center for Soil and Agroclimate Research in Bogor. For each location it equals wet months minus dry months, where a month is rated as "wet" if its average precipitation is over 200 mm and "dry" if that average is under 100 mm. Implicitly, therefore, a wet month is treated as a month with full irrigation. The value of net wet months ranged from -2 months to 9 months.

The soil chemistry variables are average values for each district as predicted by the regressions discussed in chapter 7 and then adjusted as described in note 8 to this chapter.

lower. The low intensity of inputs, even beyond the inputs measured here, depresses the yield per harvested area. By contrast, on Java, where settlement is of longer standing and cultivated lands are harvested often, the seemingly positive effect of extra arable on harvest productivity in some districts is actually an effect of less-intense cropping in those districts, allowing more selectivity concerning which plots to work more intensively.[6]

Water supply and proximity to large nonagricultural populations also have understandable effects on agricultural productivity. Water supply makes a significant difference among the outer islands, where it varies greatly between the wet northwest and such drier areas as Nusa Tenggara and southern Sulawesi. The quadratic terms for water supply imply that, for the generally unirrigated outer islands, the optimal rainfall would still be quite high: not as high as the Sumatra average, but slightly above the averages for Kalimantan and Java. For Java, rainfall was less of a constraint on productivity, partly because our 1990 data year was not an El Niño drought year. Water supply was insignificant as a variable separating the agricultural fortunes of districts within Java. The nonagricultural population density, which approach A takes to be an environmental variable reflecting market access and hidden inputs for agriculture, was highly significant within Java and in the all-Indonesia results. Being close to dense urban markets clearly raised productivity in Java. Indeed, this variable offered the most powerful explanation of differences in agricultural productivity among the districts of Java. No such effect showed up in the less densely settled outer islands.

Competing against all these influences in approach A's fuller model, soil chemistry appears to show some effects on agricultural productivity in the first panel of table 8.1. For Indonesia as a whole both soil organic matter and potassium (total K_2O content) are apparently positive influences on productivity.[7] Organic matter seems to play a role mainly within Java and in the implied contrast between Java and the outer islands. The total soil potassium endowment seems to have a high marginal productivity in the outer islands. Each of these two results implies that a soil nutrient has more effect where its endowment is poorer: organic matter and nitrogen make a visible difference in Java, where their levels are lower on the average, whereas potassium is significant mainly on the outer islands, which are very poor in soil potassium. Phosphorus (total P_2O_5 content) and acidity, by contrast, fail to show up as major determinants of interdistrict differences in productivity.

These soil effects from approach A could, by themselves, tell an important story about the agricultural consequences of Indonesia's soil changes since 1940. The importance of organic matter, on Java and across the whole archipelago, would suggest that this component of Indonesia's total investment in soil quality worsened significantly between 1940 and 1970, then improved from 1970 to 1990. The importance of potassium suggests that soil quality has advanced greatly on Java's cultivated soils but deteriorated on the outer-island soils, at least on those islands where soil samples were taken. The next task would be to quantify the net change in overall soil quality arising from these different negative and positive influences.

First, however, let us consider an alternative view of soil's importance using approach B. It may be wrong to use approach A to search for the effects of soil while holding so many other influences constant. In a market economy like Indonesia's, the other things should probably not be held constant. Labor and capital and fertilizer are not exogenous supply side forces for each separate district in Indonesia, as mentioned in connection with figure 8.1. If mobile inputs tend to move to those districts where the soil and climate are better for agriculture, the regressions of approach A may mistakenly credit these inputs with productive effects that were really caused by prior differences in soil and climate. Suppose, for example, that a district whose soils are well endowed with phosphorus tends to attract labor and capital to that district. Then some part of the extra labor and fertilizer inputs in that district are really due to the phosphorus content of its soils. A regression equation like Equations (1)–(3) in the first panel of table 8.1 may be unable to distinguish between a soil effect and the effect of extra labor and capital, if soil determines where the labor and capital go.

If we follow approach B's strategy and explain agricultural productivity *only* in terms of soil and climate, we may avoid the danger of underestimating the true causal effect of soil conditions. The second panel of table 8.1 presents the reduced-form results envisioned by approach B. These results reaffirm the importance of soil and climate, with a couple of new twists. Soil chemistry matters more on the outer islands, and in the contrast between the outer islands and Java, than it matters within Java, just as we found in the results of approach A. Indeed, in this case soil and climate have so little overall impact on differences in agricultural productivity within Java that the results of equation (5), which involved the same variables as (4) or (6) but dealt with Java specifically, were statistically insignificant.

On the outer islands, and especially for the nation as a whole, the soil characteristics that stand out as affecting output are phosphorus and potassium, versus that combination of organic matter and potassium in approach A's results. That phosphorus shows an effect in this reduced-form equation, but not when other variables are included, may be due to what the last paragraph envisioned about phosphorus: Perhaps soil better endowed with phosphorus attracts labor and capital to a district, so that the usual kind of production function equation hides the full effect of having more phosphorus-rich soils. Potassium continued in approach B to have a large predicted effect on agricultural productivity, albeit with a wide confidence interval. Soil potassium mattered more on the outer islands, where it is low on the average, than in Java. This spatial pattern accords with soil scientists' general understanding that soil potassium is a key constraint on plant growth at low levels but stops contributing at high levels (e.g., Brady 1990, p. 371). Organic matter (and nitrogen) failed to show any positive productivity effect in approach B. Perhaps for both parts of Indonesia, as for both parts of China, soil organic matter has no visible impact on productivity because it is so easily augmented with nitrate fertilizers, a point we return to in the next section.

Thus far, it appears that some soil characteristics matter for agricultural productivity and others fail to show any effect. To recapitulate, we seem to have found these elasticities of effect of soil chemistry on agricultural productivity:

• Raising soil *potassium* (total K_2O content) by 10 percent would raise agricultural output by about 1.25 percent for Indonesia as a whole, or by about 1.50 percent on the outer islands, with no clear effect in Java.

• Raising soil *phosphorus* (total P_2O_5 content) by 10 percent would raise output 1.64 percent on the outer islands, *if* one puts one's faith in the results of the simplified approach B. In the fuller specification of approach A, on the other hand, any effect of soil phosphorus is hidden behind other forces.

• Raising soil *organic matter* (and nitrogen) might raise output, if one accepts approach A's assumption that soil chemistry does not affect other input levels. In this case, it looks as though an extra 10 percent of soil organic matter would raise output by 1.18 percent for the whole nation, 1.26 percent for Java, and only 0.42 percent for the outer islands. But the range of possible error is wide enough in this case that the true effect might be zero.

• Soil *acidity* fails to show any significant effect on agricultural output in these results, unlike its effect in south China.

Before turning to the overall productivity consequences of soil conditions, however, we must note another possible channel for those consequences.

Fertilizer as a Substitute for Soil Quality

Fertilizer is itself a soil amendment, at least for the short run. Any pursuit of the productive role of soil quality therefore faces the following question: Does fertilizer act as a substitute for soil quality, a cure for poor soil? To the extent that fertilizer can cheaply replace any extra nutrients missing from the soil endowment, the importance of soil endowment as an input to agricultural productivity dwindles away.

Such a substitution possibility emerged from the exploration of China's agriculture in chapters 5 and 6, where it seemed that one particular set of soil characteristics looked surprisingly unimportant in the contrast of China's counties in the 1980s. Soil organic matter and soil nitrogen together failed to make a clear contribution to productivity in 1980s China. Although we know that soil organic matter and nitrogen are crucial to plant growth in the larger sense—no nitrogen or organic matter, no crops—it is possible that the relevant margin, the application of quick-acting fertilizer can reverse some losses in endowments of these nutrients. The evidence for this conclusion came only from cross-county regressions, however, and there is still reason to ask whether it is plausible that soil organic matter and nitrogen had little marginal effect on yields, being replaced in poorly endowed areas by nitrate fertilizers, whereas soil phosphorus and potassium continued to be important.

The Indonesian results for 1990, like the 1980s China results, hint that nitrogen delivered in fertilizer can be a significant substitute for soil nitrogen and organic matter. We are helped in our analysis of this possibility by the fact that recent Indonesian data report not just tonnages of all fertilizer, as in China's county data, but the nitrogen content, phosphorus content, and potassium content separately for government-subsidized chemical fertilizer use by district.

Indonesian district data offer a clue supporting the belief that nitrogen content from fertilizer can substitute relatively easily for soil nitrogen. Separate regressions for the 173 districts, not tabulated here, have cast the nitrogen content, phosphorus content, and potassium content of fertilizers in the role of dependent variables, to be explained by the same nonfertilizer variables as those that appear in table 8.1. These fertilizer content regressions show that nitrogen content differs from phosphorus content or potassium content in the way suggested by the less direct tests for China (approach A). Farms and estates apply more nitrogen content in those districts where the soil is poorer in nitrogen, other things equal. This suggests that Indonesian agricultural practice not only can, but actually does, substitute quick-release nitrate fertilizers, such as urea, for soil nitrogen.[8] By contrast, there was no such response of phosphorus or potassium fertilizer delivery to soil phosphorus or soil potassium, respectively. Thus far the direct contribution of soil phosphorus and potassium to productivity, with little offsetting substitution, seems confirmed.

Although this clue cannot be conclusive by itself, it does suggest that substituting quick-release nitrogen fertilizer can eclipse the role of soil organic matter and nitrogen in agricultural productivity and that this substitution option relates more easily to nitrogen than to phosphorus or potassium.

Indonesia's Net Investment in Soils, 1940–1990

Changes in Soil Chemical Quality

Knowing roughly how much an extra unit of each soil characteristic should affect agricultural output makes it possible to estimate the overall consequences of the whole historical movement of average soil characteristics. The unit impact of each soil characteristic implied by table 8.1 can play the role of its price, telling us what each unit of change in that soil chemistry characteristic is "worth" in terms of the extra agricultural production it engenders. Multiplying this by the historical soil change in that soil characteristic and adding up these terms for all characteristics tells us the overall productive consequence of the set of soil changes observed. This gives us a rough idea of the net historical change in Indonesia's soil quality, just as a similar calculation in chapter 6 did for China.

Table 8.2 sketches the changes in soil productivity implied by summing the estimated productivity effects of changes in organic matter, nitrogen, P_2O_5, K_2O, and acidity, first over a half century and then between islands at a specific point in time.[9] The results for all Indonesia at the top of the table are summed from the separate results for Java-Madura and for the outer islands.

On these measures the soil chemical quality[10] of Indonesia's cultivated lands declined over the half century 1940–1990, either by 5.5 percent of its initial value, if one accepts the approach A estimates, or by 4.0 percent, if one accepts the approach B estimates. Either figure is, of course, a very rough estimate. Why should there have been a net decline in soil chemical quality for Indonesia as a whole over the period? Continuing down table 8.2, we see that the overall decline for the whole archipelago results from that in the outer islands, since the estimates for Java-Madura allow for a possible rise in the productive quality of topsoil chemistry. For Java-Madura, topsoil chemical quality either dropped by almost 3 percent between 1940 and 1955, with partial recovery thereafter, if one prefers Approach A's emphasis on the productivity of soil organic matter and nitrogen, or it rose by almost 10 percent between 1940 and 1990, if one accepts approach B's emphasis on the buildup of P_2O_5 in Java-Madura.

The decline in topsoil chemical quality on the outer islands over the 1940–1990 half century was perhaps 11.8 percent or perhaps 17.7 percent, depending on one's choice between the soil chemical quality indexes implied by approach A and approach B. In either approach, most or all of the net decline appeared after 1960. This loss of soil chemical quality on the outer islands was sufficient to bring about that overall national loss of 4.0–5.5 percent, even after allowing for a possible gain in soil quality on Java and Madura.

Why did the productivity of outer-island soils decline in 1940–1990? The proximate source of the decline is the post-1960 decrease in levels of topsoil nutrients: soil potassium, phosphorus, organic matter, and nitrogen. This apparent decrease could have resulted from three forces discussed in chapter 7: (1) a shift to historically more marginal soils, (2) continuing human-induced soil degradation, and (3) a shift to younger (more recently cleared and cultivated) soils. As argued there, the agricultural history of the outer islands suggests that the third force was strongest. Now that the decrease in key nutrients has been converted into its implied productivity consequences, the main story seems to be that the soil chemical quality of cultivated lands of the outer islands deteriorated on the average because a rising share of the cultivated area, especially after 1970, were recently cleared lands. These recently cleared lands had lost nutrients, especially organic matter and potassium, in the clearing process but had not yet received additional nutrients from humans.

If this interpretation is correct, then the soil chemical quality decline on the outer islands need not continue in the long run. To the extent that it has resulted from recentness of settlement and cultivation, it will reverse itself as soon as the *rate* of expansion of cultivated area *slows down,* an event that will come much sooner than the time when the expansion of cultivation stops. The slower the later rate of growth of outer-island cultivated area, the closer the outer island soils will come to the quality trend shown in regions where cultivated area has stopped expanding. Both on area-stagnant Java and in area-stagnant China, a modest net rise in soil quality has set in.

That Java-Madura offers a hint about the future of agricultural soils on the outer islands is underlined by table 8.2's final calculation, the productivity difference between the soil chemical characteristics of these two parts of Indonesia. By 1990 the cultivated soils of Java and Madura had become inherently more productive, even aside from their better access to markets. The extra intrinsic productivity resulted from the wide gap in the topsoil endowments of total potassium and total phosphorus. Even granting that most of the buildup of P_2O_5 and K_2O takes forms that come available for plant growth only very slowly, there is reason to accept the regression results that these are endowments that raise productivity year after year. The important point about these extra endowments is that they were built up by humans on long-settled, now-crowded soils that were not always so productive.

Table 8.2
Soil chemistry quality and quantity of Indonesia's cultivated lands, 1940–1990

All Indonesia	1940	1955	1960	1970	1980	1990
Soil chemical quality index[a]						
Approach A	105.79	104.86	106.36	101.77	100.87	100
Approach B	104.16	103.38	102.85	103.16	101.46	100
Total cultivated (farm + estate) area						
Total area	18,109	19,178	19,842	21,068	27,322	38,024
Total area, Java equivalents	14,896	15,671	16,124	16,878	20,803	28,013
Quality-adjusted area at 1990 Java quality						
Approach A	15,757	16,432	17,149	17,177	20,983	28,013
Approach B	15,515	16,202	16,584	17,412	2,1107	28,013
Java-Madura	1,940	1,955	1,960	1,970	1,980	1,990
Soil chemical quality effects, approach A						
Average organic matter[b]	3.83	3.04	2.77	2.34	2.45	2.57
Organic matter effect	1.05	1.02	1.01	0.99	0.99	1.00
Average K_2O	29.78	29.49	29.04	32.58	34.71	37.19
K_2O effect	0.97	0.97	0.97	0.98	0.99	1.00
Implied soil quality index	101.9	99.1	97.8	97.3	98.6	100.0
Soil chemical quality effects, approach B						
Average P_2O_5	45.31	67.42	52.27	62.40	67.35	75.14
P_2O_5 effect	0.93	0.98	0.95	0.97	0.98	1.00
Average K_2O	29.78	29.49	29.04	32.58	34.71	37.19

Table 8.2
(continued)

Java-Madura, continued	1940	1955	1960	1970	1980	1990
K_2O effect	0.97	0.97	0.97	0.98	0.99	1.00
Implied soil quality index	90.3	95.7	92.0	95.7	97.6	100.0
Total cultivated (farm + estate) area	8,754	8,969	9,019	8,869	8,343	8,881
Quality-adjusted area at 1990 Java quality						
Approach A	8,922.7	8,884.1	8,821.1	8,627.6	8,224.8	8,881
Approach B	7,906.7	8,579.3	8,293.8	8,490.5	8,140.3	8,881

Outer islands	1940	1955	1960	1970	1980	1990
	1,940	1,955	1,960	1,970	1,980	1,990
Soil chemical quality effects, approach A						
Average organic matter[c]		3.60	4.37	3.01	3.25	3.51
Organic matter effect[d]		1.00	1.00	1.03	0.98	0.99
Average K_2O[c]	21.45	21.54	24.85	16.59	12.21	8.99
K_2O effect	1.12	1.12	1.14	1.08	1.04	1.00
Implied soil quality index	111.8	111.9	116.5	106.0	103.0	100.0
Soil chemical quality effects, approach B						
Average P_2O_5[c]	27.49	18.58	18.43	22.27	20.39	18.77
P_2O_5 effect	1.06	1.00	1.00	1.03	1.01	1.00
Average K_2O[c]	21.45	21.54	24.85	16.59	12.21	8.99
K_2O effect	1.11	1.11	1.13	1.08	1.04	1.00

Table 8.2
(continued)

Outer islands, continued	1940	1955	1960	1970	1980	1990
Total cultivated (farm + estate) area	9,355	10,209	10,823	12,199	18,979	29,143
Quality-adjusted area						
At 1990 outer islands quality, approach A	10,463	11,424	12,614	12,937	19,544	29,143
At 1990 outer islands quality, approach B	11,010	11,346	12,227	13,486	19,947	29,143
At 1990 Java quality, approach A	6,869	7,500	8,281.2	8,492.9	12,830	19,132
At 1990 Java quality, approach B	7,227.8	7,448.9	8,026.9	8,853.7	13,095	19,132

Java-Madura vs. outer islands, 1990	Java	Outer islands	% advantage for Java
Soil chemical quality effects, approach A			
Average organic matter	2.57	3.51	
Organic matter effect	0.96	1.00	-3.6
Average K_2O	37.2	9.0	
K_2O effect	1.19	1.00	19.5
Implied soil quality index	115.2	100.0	15.2
Soil chemical quality effects, approach B			
Average P_2O_5	75.1	18.8	
P_2O_5 effect	1.23	1.00	22.7
Average K_2O	37.2	9.0	
K_2O effect	1.19	1.00	19.0
Implied soil quality index	146.0	100.0	46.0

Table 8.2
(continued)

Notes: All areas are in 1,000s of hectares.
a Geometric average of Java and outer islands.
b Extrapolate to 1940 from 1960 using tree crop sites.
c Tree crop lands only.
d Assume 1940 index = 1955 index.

Cultivated land is approximated here by arable land of types that are oriented toward current agricultural use, as estimated in Van der Eng 1996 App. 4. It consists of total farm area and estates. In 1990 it compared with total arable and total harvested area as follows:

	Indonesia	Java	Outer Island
All arable area	40,888	8,922	31,966
Farms and estates	38,024	8,881	29,143
Harvested area	29,433	11,876	17,557
(Farms and estates)/arable	0.930	0.995	0.912

Values for each hectare of outer island area can be converted into values for a Java hectare using values per hectare of (farm + estate) in 1990:

	Indonesia	Java	Outer Island
Gross agricultural product	34,456	1,7119	17,337
Farms and estates	38,024	8881	29,143
Product/(farm + estates)	906.2	1,927.6	594.9

Each outer island hectare of (farm + estate) is this assumed to be worth $(594.9/1927.6) = .3086$ of a Java hectare. This procedure converts non-Java values into Java equivalents by using output values, not soil chemistry qualities.

The Agricultural Land Endowment, 1940–1990

The net change in Indonesia's stock of agricultural land can be judged by multiplying the land's average quality by its quantity (in hectares), as shown in table 8.2, which introduces rough estimates of the cultivated areas on all the islands, both with and without adjustment for the fact that a hectare of land cultivated on the outer islands generates less output value than a hectare on Java or Madura.

The cultivated area of Indonesia as a whole more than doubled between 1940 and 1990. Almost this entire increase came on the outer islands, with practically no expansion of cultivation on already-crowded Java. This expansion so eclipses the 4.0–5.5 percent decline in soil quality as to result in an 80 percent increase in the nation's agricultural land stock. Although the average quality of Indonesia's agricultural lands may have declined over the half century studied, the nation clearly did not run out of cultivable lands or even experience a decline in its endowment of such lands, anytime up to 1990. On the outer islands, where the decline in soil quality was concentrated, the increase in cultivable area was even faster. The number of cultivable hectares more than tripled, resulting in a near-tripling of cultivated land stock when the effects of land quality and quantity are multiplied together.

Do Soil Changes Explain the Productivity Changes?

The rough estimates of Indonesia's soil quality and quantity can be viewed from different perspectives. If soil quality declined in 1940–1990, was this decline big or small? When compared with the increase in land area, it looked small. Yet one could still seek to discount the increase in land area as something unsustainable in the long run and focus on the change in quality alone. It is reasonable to concentrate on per hectare or per person magnitudes as a guide to where Indonesia's management of its soil needs to be headed. How large or small does the 4–6 percent drop in average soil quality between 1940 and 1990 appear in relation to trends in Indonesia's ability to produce agricultural goods?

Some initial perspectives on this question are offered by table 8.3's comparisons of the soil chemistry effect with the overall changes in land productivity and labor productivity during the period 1940–1990. For the nation as a whole, land and labor productivity were both roughly

Table 8.3
Role of soil chemistry in accounting for agricultural productivity differences, Java-Madura and the outer islands, 1940–1990

A. A half century of agricultural growth, 1940–1990

	1940 level	1970 level	1990 level	Percentage changes		
				1940–1970	1970–1990	1940–1990
All Indonesia						
Gross value of production (billions of 1990 Rupiahs)	11,712	14,431	34,444	23.2	138.7	194.1
Harvested area (thousands of hectares)	16,521	19,105	26,990	15.6	41.3	63.4
Agricultural employment (millions)	19.94	26.37	35.75	32.2	35.6	79.3
Land productivity = Production (billions of Rupiahs) per hectare	709	755	1,276	6.5	69.0	80.0
Labor productivity = Production (thousands of Rupiahs) per person	587	547	963	-6.8	76.1	64.0
Output/hectare due to soil chemistry quality (Approach A)	105.8	101.8	100.0	-3.8	-1.7	-5.5
Output/hectare due to soil chemistry quality (Approach B)	104.2	103.2	100.0	-1.0	-3.1	-4.0
Java-Madura						
Gross value of production (billions of 1990 Rupiahs)	7,437	8,294	17,107	11.5	106.3	130.0
Harvested area (thousands of hectares)	9,169	9,971	11,553	8.7	15.9	26.0
Agricultural employment (millions)	13.21	16.07	18.48	21.7	15.0	39.9

Table 8.3
(continued)

	1940 level	1970 level	1990 level	Percentage changes		
				1940–1970	1970–1990	1940–1990
Land productivity = Production (billions of Rupiahs) per hectare	811	832	1,481	2.5	78.0	82.5
Labor productivity = Production (thousands of Rupiahs) per person	563	516	926	-8.3	79.4	64.4
Output/hectare due to soil chemistry quality (Approach A)	101.9	97.3	100.0	-4.6	2.8	-1.9
Output/hectare due to soil chemistry quality (Approach B)	90.3	95.7	100.0	6.0	4.5	10.7
Outer islands						
Gross value of production (billions of 1990 Rupiahs)	4,275	6,137	17,337	43.6	182.5	305.6
Harvested area (thousands of hectares)	7352	9134	15437	24.2	69.0	110.0
Agricultural employment (millions)	6.73	10.30	17.27	53.0	67.7	156.6
Land productivity—Production (billions of Rupiahs) per hectare	581	672	1,123	15.6	67.1	93.2
Labor productivity—Production (thousands of Rupiahs) per person	635	596	1004	-6.2	68.5	58.1
Output/hectare due to soil chemistry quality (Approach A)	111.8	106.0	100.0	-5.2	-5.7	-10.6
Output/hectare due to soil chemistry quality (Approach B)	117.7	110.6	100.0	-6.1	-9.5	-15.0

Table 8.3
(continued)

B. Java-Madura versus the outer islands in 1990

	Java-Madura	Outer islands	% difference, Java–Outer islands
Gross value of production (billions of 1990 Rupiahs)	17,107	17,337	−1.3
Harvested area (thousands of hectares)	11,553	15,437	−25.2
Agricultural employment (millions)	18.48	17.27	7.0
Land productivity—Production (billions of Rupiahs) per hectare	1,481	1,123	31.8
Labor productivity—Production (thousands of Rupiahs) per person	926	1,004	−7.8
Output/hectare due to soil chemistry quality (Approach A)	115.2	100.0	15.2
Output/hectare due to soil chemistry quality (Approach B)	146.0	100.0	46.0

Source: The Appendices in van der Eng 1996 were used for all nonsoil aggregates.
Notes: Soil measures are based on this study's soil samples and the regression results in table 8.1. Figures for gross value of production for 1940 and 1970 are based on the assumption of a constant share of real value added in the gross value of real output.

stagnant between 1940 and 1970, then jumped by more than two thirds between 1970 and 1990. Against the backdrops of the large gains in productivity in the 1970s and 1980s, the accompanying loss in soil quality looks small. Taken at face value as a measure of lost productivity, the soil chemical quality loss was 1.7–3.1 percent over these two decades, and again, 4.0–5.5 percent over the whole half century. Only to this extent could the decline in soil chemical quality be said to have reduced the land productivity gain to 60 percent in the 1970s and 1980s and to 80 percent over the half century, and to have reduced the labor productivity gain—or, roughly speaking, the rural living standard—to 76 percent for the 1970s and 1980s or 64 percent for the whole half century. For both Java-Madura and the outer islands, the productivity gains are much larger than the changes in soil quality between 1940 and 1990.

In one perspective, however, differences in soil quality loom large relative to overall productivity differences, and this perspective returns us to some basic thoughts about growth and soil quality. As shown by the last panel in table 8.3, Java-Madura's productivity advantage due to topsoil chemical quality alone was large enough to be roughly comparable to Java-Madura's overall advantage in agricultural output per harvested hectare. It was much larger than the contrast in output per laborer, which actually favored the outer islands slightly. Here, then, is a way in which differences in soil quality loom large. The outlook is not so negative, however, regarding the process of human settlement, as soil quality is noticeably higher where a dense population has worked the land intensively for decades or centuries.

IV

Conclusions and Implications

9

What Have We Done to the Land?

Drawing conclusions from limited and novel data sets is dangerous but worthwhile. Like the Hubble telescope before its expensive repair, the new data sets can give only a blurry view from a superior vantage point. Yet even a strong dose of caution does not prevent us from reaching clear conclusions. So broad and sweeping are the past assertions about soil trends that even imperfect data can resolve several of the biggest questions for China and Indonesia. Replicating such soil history studies, even for just one or two other countries, can supplement what we are learning from the continuing official efforts to track soil trends into the twenty-first century.

This chapter takes stock of the major results concerning soil trends that now show up quite clearly when the available data are analyzed, even though the data are still too limited and imperfect for a closer look at the underlying soil dynamics.

On Trends in Soil Characteristics

Changes in Organic Matter and Nitrogen Levels

The intensification of agriculture does not enhance soil nitrogen and organic matter and often depletes it. Total nitrogen and organic matter surely decline when previous forest or grasslands are first cleared and cultivated. The decline may even continue thereafter, as cultivation intensifies.

The timing of the apparent declines in organic matter and nitrogen was different in the two countries studied here. For Indonesia, the decline was probably concentrated in the period before 1970, though data limitations mean that this conclusion rests mainly on the movement of levels of organic matter in Java. For China, the evidence for a decline is clearer for the 1950s–1980s period than for the earlier 1930s–1950s period.

Changes in Phosphorus and Potassium Levels

Unlike those of nitrogen and organic matter, the total soil endowments of phosphorus and potassium have increased in the long-cultivated regions of both countries. For China, the only countercurrent to this trend is that total phosphorus appears to have declined after the 1950s on the Huang-Huai-Hai plain, in contrast to the pattern in all other regions. With respect to Indonesia, both phosphorus and potassium increased markedly over the half century from 1940 to 1990 in Java. The general decline of these two nutrients in the outer islands seems to be due to the acceleration of the settlement process, as explained in chapter 7.

Changes in pH

Soil alkalinity did not clearly worsen in north China between the 1930s and the 1980s. Rather, it remained serious in the northwest, while China won victories against alkalinity and salinity on the Jiangsu and Bohai coasts.

The threat of acidity in subtropical and tropical areas remains, but without any monotonic trend over time. South China experienced some abatement of acidity between the 1930s and the 1950s, but acidity increased from the 1950s to the 1980s, leaving no clear overall trend. Soils in Indonesia showed was no trend in acidity (and possibly some improvement) before 1970 and some worsening after 1970. Unless acidity is abated there, the concentration on acid-tolerant rice paddy and tree crops will continue.

On Their Causes

What human-induced processes of soil change were dominant causes of the observed geography and history of soil change in China and Indonesia?

Changes in Soil pH

Water control seems to have been the key to improvements in soil pH in China, helped by large-scale investments in levees, drains, and irrigation ditches. Such measures won victories over alkalinity in the Bohai and

Jiangsu coasts, and kept alkalinity from worsening in the northwest. In Indonesia, the development of liming techniques appropriate to Indonesia's climate and soils aided efforts at keeping acidity under control.

Changes in Macronutrient Balances

We have used contrasts in the temporal and spatial patterns of soil chemistry to make educated preliminary conjectures about why soil nutrient levels changed as they did in the areas and time periods studied. For organic matter and the NPK macronutrients, the task—a difficult one—is to work out how the whole multivariate balance of soil gains and losses for each nutrient has varied over time and space. Even if we know background parameter values for natural rates of gain or loss, such as nitrogen fixation or nitrification losses, key elements of the balance are still elusive, particularly the unmeasured human applications of organic fertilizer not purchased from the government or by large firms.

What we know about the natural processes does seem consistent with the potentially puzzling difference between the net depletion of soil nitrogen and the net buildup of total phosphorus and total potassium. The nitrogen cycle involves more natural volatility and more variability in available soil nitrogen than the corresponding cycles for potassium and phosphorus, so that both accumulation and losses occur at a faster rate than the accumulation and losses of (either total or available) phosphorus and potassium. The total nitrogen supplied by soil even has a pronounced seasonal pattern in many climates (Institute of Soil Science 1990, chap. 29; Brady 1990, chaps. 11–12). Heavy applications of nitrogen-rich quick-release chemical fertilizers, especially urea, can offset the faster natural loss rates for nitrogen. For China, at least, the nitrogen constraint is both frequently binding and relatively responsive to fertilizer in the short run.

Within this complex natural balance, variations in fertilizer applications and in crop uptake seem to have been the main factors shaping geographic patterns and trends in soil nitrogen, phosphorus, and potassium. For example, the patterns in China relating to soil organic matter and nitrogen seem to have been more favorable wherever and whenever cropping was less intense (toward the west, and between the 1930s and 1950s) than in those places and times in which it was more intense (toward the east, and between the 1950s and 1980s). This suggests, in the

Chinese setting, that differences in agricultural intensity had more effect on soil nutrients than did differences in the rate of fertilizer inputs.

The future trend of macronutrient balances may be both favorable and unfavorable for China, and to a lesser extent for Indonesia, as the average incomes grow. The trend may be favorable in that chemical fertilizer supplies should go on improving, especially in China, where fertilizer has been peculiarly expensive relative to the prices of the crops it produces (Stone 1986, 1988). This should continue to raise phosphorus and potassium levels in the soil while supplying more nitrogen for plant growth without enhancing total soil nitrogen much.[1] The trend is likely to be unfavorable in terms of the supply of organic fertilizers. The trends toward larger cities might seem to favor the collection and application of such fertilizers, but the exodus of labor from agricultural areas and the rapid rise in China's real wage rates will weigh heavily against this. So labor intensive is the organic-waste sector, and so unattractive in its work regime, that rising wages should bring major cuts in the supply of organic wastes, especially in China. How these favorable and the unfavorable fertilizer trends will balance out remains to be seen.

The Case against the Dominance of Erosion

The evidence in this study contradicts the traditional emphasis on erosion as the number-one soil degradation threat, which has promoted three ideas at once: (1) that erosion is the dominant source of soil degradation; (2) that it is mainly human induced; and (3) that it is accelerating in tandem with the intensity of human agriculture. Several kinds of evidence from China and Indonesia contradict all three of these traditional beliefs about erosion:

• The accumulation of total phosphorus and potassium in both countries contradicts the tendency of erosion to draw down all macronutrients in the remaining topsoil.

• The spatial locus of China's decline in soil nitrogen does not match erosion's fingerprints. Nitrogen has not declined most, and perhaps has not declined at all, in the areas that earlier writings considered the most prone to erosion. Rather, the declines are clearest in the Huang-Huai-Hai plain and other eastern areas not traditionally thought of as sites of high erosion.

• We now know that in the most erosion-prone areas, particularly in the loess plateau and arid fringes of China's northwest, the cultivated areas have been expanding, not declining, over the half century ending around 1990. The desert and the gullies have not advanced onto farmlands as much as the farmlands have advanced to the edges of the deserts and gullies (as noted again in the next section).

• What appears at first glance to be a confirmation of erosion's importance in fact fails to deliver that confirmation when examined more closely. In both countries, the topsoil layer, called the A horizon in current taxonomy, appears to have become thinner, especially before the 1950s, and such thinning would normally confirm a hypothesis of erosion. Yet this apparent early thinning may have been due only to a switch in the definitions of topsoil layers rather than to any actual physical thinning of the topsoil layer.

If erosion processes have attracted more than their share of attention relative to other processes, as the evidence seems to suggest, why did this happen? How did it come about that researchers and the public focused almost exclusively on erosion? Part of the answer may be that concerns about erosion can be developed faster, more dramatically, and with greater implications for social intervention than concerns about other types of soil degradation. The initial evidence of erosion is easy to deliver: Just measure river siltation rates for each watershed and add photographs of gullies and sand dunes to dramatize the problem's potential for inflicting lasting damage. It is much more difficult, on the other hand, to probe the agronomy and soil science of nutrient imbalance, which requires patient soil and crop sampling over a period of years. And by focusing on erosion in the large, rather than on an individual farmer's plowed fields in the off-season, the writers dramatizing erosion dangers have also been able to focus our attention on erosion of a sort that cannot be checked without large-scale collective actions, pushing the erosion issue toward the center of the government policy spotlight.

On Changes in Cultivated Land Area

In both China and Indonesia, total cultivated area has risen, though more slowly than overall human population. For China the expansion of total cultivated area has been revealed only in the 1990s, after many writers had concluded that China's agricultural land base was shrinking. Only with the help of satellite reconnaissance have we discovered how much land China's farm population is actually working.

A first step toward understanding the causes of the increase in land area under cultivation is to weigh the importance of farmland losses to construction. Large as these land losses have loomed in the global imagination, the actual information available suggests that for China they are quite limited. True, the lands lost to construction are of higher than average soil quality, but the ratio of their average soil quality to the grand average quality of agricultural soils is probably less than two to one. Even after allowing for this quality difference, the net loss of farmland to construction is still quite slow, though it is accelerating. Against even this small loss should be weighed the unknown gain in the quality of the soils remaining in agricultural areas near the spreading cities, a gain caused by the tendency for a larger nonagricultural population to bring better supplies of fertilizer and greater incentive for farmers to invest in soil improvement.

For Indonesia, virtually none of the gains in cultivated area since 1970 have occurred on Java. Rather the gains in cultivated area consist of settlements on the outer islands, a process that will eventually slow down.

Consequences

Which Soil Characteristics Seemed to Matter at the Margin?
The historical stability or decline in soil organic matter and total nitrogen has had no clear effect on agricultural yields, as far as we can tell from statistical explorations using data from both countries. The main reason for this lack of clear effect appears to be that quick-release nitrogen fertilizers can compensate for any losses of soil nitrogen within the same crop season. In this respect, the cross-sectional evidence from China and Indonesia seems to resemble the findings, from the classic long-term experimental plots, that reapplying standard inorganic fertilizers promptly redresses deficiencies in soil organic matter and nitrogen. For irrigated lowland rice agriculture, an additional reason for this lack of effect is that total soil organic matter and total nitrogen are both stable and lack a strong marginal influence on available nitrogen.

The level of total potassium seems to have mattered to yields, both in south China and throughout Indonesia. Even though only a small fraction of total potassium becomes available for plant growth in a given year, sufficient total-soil potassium has built up to raise yields noticeably.

The effects on yields of other soil characteristics, beyond potassium, varied between north and south China and between Java and the other islands of Indonesia. In some of the China results, acidity and alkalinity affected yields, and in some of the Indonesia results the total phosphorus endowment made a significant difference. The magnitude of these influences remains uncertain, however.

The Net Farmland Investment in These Two Countries So Far
Weighing each soil characteristic by its marginal impact on agricultural yields and multiplying these marginal impacts by the stock of cultivated land produces a measure of the endowment of agricultural soil capital for a given area. Changes in this endowment over time represent the area's net investment in productive soils.

In the historical interaction between changes in soil quality and changes in farmland area, the growth of farmland turns out to dominate the net soil investments of both China and Indonesia. Table 9.1 underlines this result. The changes in topsoil quality (panel A) since the 1930s have been modest and probably not negative, the exception being the deterioration in the average quality of cultivated soils on Indonesia's outer islands since 1960. The national rates of expansion of cultivated area (panel B) have been greater than any rates of improvement or degradation of the soil. Thus the product of soil chemical quality and cultivated area quantity rose over time in both countries.

Prospects for the Endowment of Farmland and Its Quality
The trends in total farmland endowments in Indonesia and China will probably not be very negative in the near future, as the prevailing pessimism has feared in the case of China. But for reasons detailed below, in the best of worlds, China could afford to cut its cultivated area, and the expansion in Indonesia's cultivated area should taper down toward zero. Although the nondecline in average soil quality suggested by the data is quite welcome, China should be prepared to accept a decline in the area

Table 9.1
Summary of quality and quantity trends in cultivated soils of China and Indonesia since the 1930s

Panel A. Topsoil quality indexes

(Jiangsu 1985 = 100)	c. 1957	c. 1985
North China, 4 regions	97.5	98.5
South China, 4 regions	79.5	95.0
All China, 8 regions	90.4	97.2

(Java-Madura 1990 = 100)	1940	1955	1960	1970	1980	1990
Java-Madura	96.1	97.4	94.9	96.5	98.1	100.0
Outer islands	75.3	73.2	75.3	71.1	68.3	65.7
All Indonesia	86.3	85.1	85.0	82.1	77.0	73.7

Panel B. Quantity alone: Cultivated land areas (millions of hectares)

	1933	1957	1978	1985	1993	1995
China	108.9	111.7	136.8	133.3	130.9	130.9

	1940	1955	1960	1970	1980	1990
Java-Madura	8.8	9.0	9.0	8.9	8.3	8.9
Outer islands	9.4	10.2	10.8	12.2	19.0	29.1
All Indonesia	18.1	19.2	19.8	21.1	27.3	38.0

Table 9.1
(continued)

Panel C. Overall agricultural land endowment—cultivated area, quality-adjusted (millions of hectares)

| | 1940 | c. 1957 | | | | c. 1985 | |
		1955	1960	1970	1980		1990
All China		104.0				133.5	
Java-Madura	8.4	8.7	8.6	8.6	8.2		8.9
Outer islands	7.2	7.6	8.3	8.7	12.9		19.1
All Indonesia	15.6	16.3	16.9	17.3	21.0		28.0

Sources: Tables 6.1, 6.4, and 8.2.
Notes: Quality indices for Indonesia are the midpoints of the approach A and approach B estimates from table 8.2. The all-Indonesia land areas consist of the lands in farms plus estates.

cultivated in food grains, as dictated by market forces. One would also hope that the expansion of Indonesia's cultivation area will come to an orderly halt. Sound development policies would bring these desirable, but seldom desired, slowdowns in cultivation while maximizing the ability of both countries to feed themselves.

This may seem odd as a summary statement, given the professed devotion of government officials, of scholars, and of journalists to the goal of maximizing the supply capacity of both countries' agriculture. Yet it follows from the logic of comparative advantage in China, and from the urgency of reversing the larger trend toward environmental degradation in Indonesia.

China has about a fifth of the world's population and less than a twelfth of its agricultural land. Given this fundamental reality, and given the demonstrated work ethic and competitive success of China's industrial export sector, China will maximize its average income *and its ability to feed itself* by shifting its labor and capital away from agricultural production. Within agriculture, it should shift further away from staple grains, which are land-intensive and less in demand as incomes rise, and toward fruits and vegetables, which are land-sparing and more in demand as incomes rise.[2] Whether China's agricultural land area actually declines depends on many things, especially government policies, but it would be no cause for alarm if the cultivated area did decline in response to a pull of labor away from agriculture and a decline in the relative scarcity of agricultural products. In the best of worlds, China's farmlands will shift from grains to a mixture of forest, grasslands, gardens, and, yes, even urban construction.[3] Each of these trends is already in motion, ushered on by improvements in China's policies and by the prolonged drift toward lower relative prices for food grains around the world.

With better institutions and policies, Java too, like China, will shift its labor and capital away from agriculture, especially away from land-intensive grains and cassava. That crowded island will have a comparative advantage in manufactures and in only selected agricultural products, such as market-oriented fruits and vegetables and tree crops. Although one imagines only stagnation and not decline in the amount of Java's cultivated area, that area will continue to drift away from staple foods when Indonesian income growth resumes.

For Indonesia's outer islands, the prospect is that the agricultural land area will go on expanding, and the average soil quality of cultivated lands may still decline a bit but should eventually rise as the age of cultivated areas increases. It is at least possible that the trend in total cultivable land endowment on the outer islands will turn negative for a while, perhaps if the observed decline in soil quality is destined to continue, since the quality of cultivated land must stop improving eventually. But as we argued in chapters 7 and 8, the process that has caused soil quality decline on the outer islands will reverse itself, and sooner than the end to land clearing. The process of soil quality decline depended on a post-1970 acceleration, not just a continuation, of settlement levels. The shift toward cultivating younger, less fertilized areas brought much of the decline. So when the rate of growth in outer-island cultivated area decelerates, the soil quality will stabilize or, more likely, rise. And for the outer islands, that deceleration in the expansion of cultivated area is to be hoped for, not just to allow the gradual revival of soil nutrients but also to slow the harmful atmospheric effects of land clearance burning and the concomitant assault on wildlife.

Will Economic Development Mine the Soil?
To underline two optimistic possible links between economic development and soil quality as a counterweight to the abundance of pessimistic scenarios, let us consider two ways in which rising incomes might improve soil quality without anybody's being conscious of such links.

Engel Effects on the Soil   Chapter 5 introduced an important possible link between economic development and soil quality. Other things equal, does the rise of average incomes that typically comes with economic development improve or worsen human treatment of the soil endowment?

To pose such a question is to revisit Engel's law. This well-supported empirical law states that a rise in any household's income lowers the share of its total expenditures accounted for by spending on food and raises the share spent on other things. Food is thus asserted to be a staple overall, where "staple" is defined as any product with an income elasticity below one, as contrasted with a "luxury," which is any product with an income elasticity above one. Tracing the implications of Engel's law at the world level, one sees that if incomes rose for the whole world, world demand would shift away from food products and world output would soon follow suit.

Next comes the potential Engel effect on the soil: If income gains mean that production shifts away from growing staple food grains and toward other activities, would this reallocate the land to less soil-degrading and more soil-conserving uses? To illustrate the possibility, let us take a simple hypothetical example, in this case an optimistic one. Suppose that poverty promoted the consumption and cultivation of wheat, whereas a rise in incomes would shift demand and cultivation toward legumes. Then income growth would have an automatically favorable effect on soil nitrogen and organic matter, as well as a broader soil-enhancing effect if prosperity also brought heavier use of fertilizers and other improvements.

Does it really work that way? All things considered, are the effects of income growth on land use such that it improves soil quality? Chapter 5 did find some limited evidence of a favorable Engel effect on soil. For both north and south China, shifting land and other resources away from the three main crop groups (grains, oils, and cotton) and into such other products as vegetables and animal products significantly raised soil organic matter and nitrogen. This resembles the favorable Engel effect that the last paragraph tried to envision. Could that be true in general? This is an area worthy of further research.[4]

Development and Soil Investment Incentives   A second basic way in which economic progress is likely to improve agricultural soils has been implicit in many authors' writings, and is independent of the new soil data presented in this book, so that it need be stated only briefly here.

A key ingredient to long-run economic development is that it reduces capital costs and clarifies property rights. To reap any gain that exists only in the future requires an investment expense in the present. The lower the real interest rate, and the less severe the rationing of capital, the more farmers and others can invest for the long term. Every behavior that can improve the land requires such waiting and would benefit from cheaper capital. So it is with investments in conservation and water control. Development makes them more profitable by cheapening capital and by giving the investor the clear title to the fruits of such investments in land.

Postscript: Trends versus Potential in Soil Management

Even if we could measure soil trends perfectly, and they turned out to be exactly as suggested in this study, finding better ways of managing the soil could still bring huge gains not shown here. "Managing the soil" here refers to a whole range of options, from soil conservation to improved fertilizer strategies and crop choices.

We know these two issues differ greatly for at least three reasons:

1. Soils vary greatly at the most local level, even within the same ordinary-sized farm plot. Soil scientists have long respected this local variability and have understood that it prevents any easy translation of regionwide conditions into local soil management prescriptions. To see how little the aggregate trends tell us about the gains and losses from specific management choices, an economist should think of an analogous contrast between average performance and the potential gains from specific investments. Even with a flat trend and modest level of overall rates of return on investments, the returns to getting a specific investment right are still huge. Returning to soil science specifically, the literature on soil conservation and management in China and Indonesia wisely spends most of its time appraising very specific local options. Examples of such localized attention to physical details of soil conservation in south China are Parham, Durana, and Hess 1993 and Wang and Gong 1998. On Indonesia, fertilizer management options are well covered by Diamond et al. 1988 and by Blair and Lefloy 1990, and upland soil conservation options by Carson 1989 and Barbier 1990.

2. Now that the estimates of China's cultivated area have been revised upward by 38 percent, our view of China's yields per cultivated hectare must be revised downward by an unknown percentage that could approach 38 percent. That being the case, China's potential for improving yields may be greater than imagined. Getting soil management right, and getting the right combinations of crop varieties and fertilizer mix, should have great potential for improving land productivity in China.

3. Large areas of Indonesia's outer islands are still in the exploitative forest mining and soil mining phase. There is great potential for them to begin approaching Java's efficiency in land use and soil management, even though Java's soils have always had somewhat greater potential than those of other islands of the archipelago.

Thus even with an average performance that is not damaging the soil increasingly over time, there is surely room for improvement, and the returns from correct soil management choices at the local level can still be great.

Still, knowing the trends delivers answers to a big question at the center the debate on human-induced soil degradation: Are humans worsening the quality of the soil by mismanaging it more and more? With the possible exception of the recently accelerated settlement of the outer islands of Indonesia, the tentative answer for these two key countries is no, *on the average,* they are not.

Appendix A: Full Soil History Regression Equations for China, 1930s–1980s

Table A.1
Determinants of soil nutrient levels in north China, 1930s–1980s (full equations)

Independent variables	(1) ln (percentage organic matter) Coeff.	$|t|$	(2) ln (percentage nitrogen) Coeff.	$|t|$	(3) ln (percentage phosphorus) Coeff.	$|t|$	(4) ln (percentage potassium) Coeff.	$|t|$
Time period and region (relative to 1980s, same region; – = positive growth up to 1980s)								
1930s, winter wheat–millet	-0.0587	(0.37)	-0.1280	(0.80)	0.1268	(1.10)	-0.057277	(0.68)
1930s, spring wheat	0.4717	(1.89)[a]	-0.2528	(0.78)				
1930s, Huang-Huai-Hai	0.1849	(0.68)	0.5521	(1.71)[b]	-0.2018	(1.63)[b]	-0.42655	(3.53)**
1930s, desert-steppe	-0.6550	(0.90)	-0.7413	(1.05)				
1950s, spring wheat	0.1422	(1.82)[a]	0.0647	(0.88)	-0.3977	(2.54)*	-0.27384	(2.57)*
1950s, winter wheat–millet	0.1401	(0.59)	0.0629	(0.36)				
1950s, Huang-Huai-Hai	0.2153	(1.15)	0.3689	(2.05)*	0.2635	(1.64)[b]	0.0515	(0.49)
1950s, desert-steppe	-0.4054	(0.86)	-0.5116	(1.17)				
Region (relative to the Huang-Huai-Hai plain, same time period)								
Spring wheat region	0.3368	(3.82)**	0.3509	(4.01)**	0.2466	(2.71)**	0.0662	(1.41)
Winter wheat–millet	0.4424	(5.71)**	0.4868	(6.72)**	0.1085	(1.76)[b]	0.1006	(2.31)*
Desert-steppe region	0.9124	(1.94)[a]	0.9706	(2.21)*	0.1238	(0.44)	0.0672	(0.38)
Northeast provinces	0.1161	(1.34)	0.0595	(0.76)	-0.1969	(1.79)[b]	0.0195	(0.46)
Density of human employment in that county in 1982								
ln (agricultural density), T	-0.1047	(4.65)**	-0.0871	(4.07)**	-0.0244	(1.17)	0.0072	(0.49)
ln (nonagricultural), T	0.0777	(3.57)**	0.0448	(2.07)[b]	0.0112	(0.53)	0.0080	(0.62)
ln (agricultural density), NL	-0.1578	(3.33)**	-0.0743	(1.66)[b]	0.0153	(0.31)	-0.0060	(0.24)

Table A.1
(continued)

| | Equation numbers and dependent variables | | | | | | |
| | (1) ln (percentage organic matter) | | (2) ln (percentage nitrogen) | | (3) ln (percentage phosphorus) | | (4) ln (percentage potassium) | |
Independent variables	Coeff.	$|t|$	Coeff.	$|t|$	Coeff.	$|t|$	Coeff.	$|t|$
ln (nonagricultural), NL	0.0729	(1.71)[b]	0.0254	(0.62)	−0.0320	(0.67)	0.0046	(0.18)
No density given	0.0498	(0.37)						
Landscape descriptions at profile site (other than tilled fields)								
Grass	0.1948	(2.52)*	0.1784	(2.44)*	−0.0721	(0.94)	0.0365	(0.78)
Marsh	−0.0653	(1.16)	−0.0382	(0.71)	0.0012	(0.02)	0.0258	(0.53)
Forest	0.4325	(5.01)**	0.2177	(2.69)**	0.0443	(0.53)	−0.0036	(0.07)
Waste	−0.1955	(2.10)*	−0.2575	(2.90)**	−0.1070	(1.10)	0.0593	(1.06)
Hill	−0.1008	(2.05)*	−0.0352	(0.76)	−0.0207	(0.42)	0.0293	(0.90)
Steep	0.1131	(1.47)	0.0172	(0.23)	−0.0852	(1.17)	0.0187	(0.44)
Mountain area	0.1586	(2.18)*	0.1023	(1.51)	−0.0420	(0.55)	0.0129	(0.30)
No landscape description given	0.3646	(2.31)*	0.1338	(0.92)	0.0599	(0.37)	0.0813	(0.98)
Parent material (other than alluvial)								
Weathered crystalline	0.1200	(1.49)	0.0441	(0.56)	0.0169	(0.20)	−0.0157	(0.36)
Weathered clastic	−0.0799	(0.96)	−0.0901	(1.10)	−0.0829	(0.92)	−0.0606	(1.20)
Weathered calcareous	0.0099	(0.11)	−0.0269	(0.30)	−0.2326	(2.34)*	0.0205	(0.33)
Diluvial	0.0633	(0.89)	−0.0065	(0.09)	−0.0538	(0.76)	−0.0517	(1.30)

Table A.1
(continued)

| | Equation numbers and dependent variables | | | | | | | |
| | (1) ln (percentage organic matter) | | (2) ln (percentage nitrogen) | | (3) ln (percentage phosphorus) | | (4) ln (percentage potassium) | |
Independent variables	Coeff.	|t|	Coeff.	|t|	Coeff.	|t|	Coeff.	|t|
Quaternary red clay	-0.3997	(3.04)**	-0.2822	(2.19)*	0.2960	(2.38)*	0.0015	(0.02)
Lacustrine deposits	0.0390	(0.35)	-0.0254	(0.25)	-0.3682	(3.21)**	0.0471	(0.61)
Coastal marine deposits	0.0753	(0.56)	-0.1058	(0.87)	-0.5033	(3.11)**	0.1780	(1.43)
Loess	-0.2235	(2.95)**	-0.1797	(2.42)*	-0.1020	(1.39)	0.0206	(0.53)
Aeolian sands	0.0695	(0.44)	0.0578	(0.36)	-0.3753	(2.15)*	-0.0641	(0.67)
Soil classifications (other than chao tu fluvo-aquic; from Agricultural Atlas)								
Yellow & torrid red	0.1041	(0.79)	-0.0728	(0.59)	-0.3510	(3.13)**	-0.1302	(1.27)
Brown earths	0.2795	(3.17)**	0.1853	(2.15)*	0.2260	(2.38)*	-0.0120	(0.22)
Dark brown forest soils	0.6881	(4.95)**	0.5248	(3.92)**	0.1713	(1.23)	0.0383	(0.57)
Gray forest, podzolic	0.8279	(4.37)**	0.5815	(3.11)**	-0.0771	(0.38)	0.1540	(1.50)
Cinnamon soils	0.0747	(0.97)	0.0740	(0.97)	-0.0822	(1.01)	-0.0776	(1.55)
Loutu, old loessial	0.0935	(0.38)	0.0962	(0.42)	0.1465	(0.63)	-0.0007	(0.00)
Gray-cinnamon forest	1.1062	(6.74)**	0.8612	(5.52)**	0.2960	(1.78)b	-0.0272	(0.33)
Heilutu, dark loessial	0.0635	(0.34)	-0.0102	(0.06)	-0.1406	(0.83)	-0.2968	(2.99)**
Chernozems	0.8638	(7.88)**	0.7592	(7.05)**	0.2792	(2.17)*	-0.0212	(0.34)
Chestnut soils	0.3718	(4.04)**	0.3819	(4.03)**	-0.1251	(1.22)	-0.0669	(1.17)
Calcic brown	-0.3619	(2.05)*	-0.2793	(1.64)b	0.1243	(0.29)	0.0546	(0.21)
Gray-desert	-0.4560	(2.97)**	-0.3284	(2.24)*	-0.0699	(0.47)	-0.0714	(0.85)

Table A.1
(continued)

| | Equation numbers and dependent variables | | | | | | | |
| | (1) ln (percentage organic matter) | | (2) ln (percentage nitrogen) | | (3) ln (percentage phosphorus) | | (4) ln (percentage potassium) | |
Independent variables	Coeff.	|t|	Coeff.	|t|	Coeff.	|t|	Coeff.	|t|
Heitu, albic	0.9843	(7.94)**	0.8486	(7.32)**	0.2498	(1.84)[a]	-0.0387	(0.65)
Calcic black	0.1716	(1.35)	0.2319	(1.93)[a]	-0.0169	(0.14)	-0.0699	(0.71)
Paddy soils	0.4590	(5.12)**	0.4187	(4.96)**	0.0087	(0.09)	0.0021	(0.04)
Saline soils	-0.4126	(4.25)**	-0.3484	(3.69)**	0.0603	(0.53)	0.0106	(0.16)
Purple soils	0.5580	(2.26)*	0.4641	(2.22)*	0.1041	(0.46)	-0.0343	(0.12)
Limestone soils	1.0713	(3.57)**	0.9171	(3.21)**	0.6240	(1.67)[b]	-0.2127	(1.62)
Yellow cult. loessial	-0.4119	(2.66)**	-0.3378	(2.26)*	0.0328	(0.21)	-0.1459	(1.66)
Aeolian sands	-0.6017	(3.27)**	-0.3427	(1.85)[a]	-0.0254	(0.12)	-0.0428	(0.38)
Subalpine & alpine	1.2035	(5.90)**	0.8313	(4.28)**	0.1274	(0.56)	0.1382	(1.27)
Alluvial, fluvisols	-0.4819	(4.55)**	-0.4291	(4.17)**	-0.0088	(0.08)	-0.0433	(0.69)
Rocky	0.1168	(1.12)	0.0908	(0.90)	-0.2367	(2.03)*	-0.0874	(1.24)
Red, volcanic	0.0112	(0.07)	0.2218	(1.45)	-0.0974	(0.63)	-0.0251	(0.28)
Precipitation (meters per year)								
Precipitation	0.1097	(0.18)	0.0772	(0.13)	0.2451	(0.38)	0.3254	(0.97)
Precipitation squared	0.0461	(0.10)	0.2039	(0.47)	0.0837	(0.18)	-0.3468	(1.33)
Other variables								
Shallow, (cm)$^{-1}$	1.4317	(3.91)**	1.0456	(2.98)**	-0.3824	(1.02)	0.0101	(1.27)
Sand %	0.0106	(1.10)	0.0138	(1.17)	0.0074	(0.62)	0.0101	(1.27)
Sand % squared	-0.0002	(3.31)**	-0.0002	(2.85)**	-0.00011767	(1.39)	-0.000025726	(0.46)
Clay %	-0.0038	(0.23)	0.0127	(0.70)	0.0069	(0.36)	0.0267	(2.05)*

Table A.1
(continued)

Independent variables	Equation numbers and dependent variables							
	(1) ln (percentage organic matter)		(2) ln (percentage nitrogen)		(3) ln (percentage phosphorus)		(4) ln (percentage potassium)	
	Coeff.	|t|	Coeff.	|t|	Coeff.	|t|	Coeff.	|t|
Clay % squared	0.000047285	(0.22)	0.0001	(0.64)	−0.0001359	(0.60)	−0.0002	(1.32)
Sand times clay	0.00016	(0.84)	0.000014112	(0.06)	−0.000085031	(0.35)	−0.0003	(2.01)*
No texture given	0.0323	(0.09)	0.2255	(0.52)	0.0613	(0.14)	0.5993	(2.01)*
Kovda's gumus, 1950s	−0.3917	(0.74)						
Source = Shaanxi 1980s							−0.1933	(2.69)**
Source = Liaoning 1980s					−0.8858	(3.59)**		
Source = Nei Mongol 1980s					−1.4764	(6.23)**		
Implicit constant	0.0561	(0.12)	−2.8839	(5.65)**	−2.8401	(5.40)**	−0.0345	(0.11)
GLS equation constant	−0.3853	(1.21)	−0.5255	(1.72)[b]	0.2648	(0.84)	−0.3073	(1.28)
Adjusted R-squared	0.6015		0.7872		0.7329		0.4664	
Dependent variable mean	0.3723		−2.5839		−3.2178		0.7654	
Standard error of estimate	0.5590		0.5313		0.5477		0.3167	
Least-squares type	GLS		GLS		GLS		GLS	
Number of observation, degrees of freedom	1,164, 1,095		1,140, 1,072		1,017, 952		751, 687	

Notes: Statistical significance (two-tailed): **1-percent level, *5-percent, [a]7-percent, [b]10-percent. For GLS equations the dependent variable and its standard error are those of the implicit equation, not the transformed equation in which all variables have been divided by a predicted standard error. T indicates that the profile itself is from a tilled field. NL indicates that no landscape description is given for this profile.

Table A.2
Determinants of pH levels in north China, 1930s–1980s (full equations)

Equation numbers and dependent variables

	(1) Northwest		(2) The Huang-Huai-Hai plain		(3) Coast		(4) Northeast provinces	
Independent variables	Coeff.	|t|	Coeff.	|t|	Coeff.	|t|	Coeff.	|t|
Time period and region (relative to 1980s, same region; − = positive growth up to 1980s)								
1930s, whole sample	-0.0181	(0.20)			0.7898	(1.98)[a]	-0.7549	(3.46)**
1930s, coast	1.0208	(1.77)[b]						
1930s, desert-steppe	0.4018	(3.55)**						
1930s, spring wheat	-0.0657	(0.66)						
1930s, winter wheat–millet	0.2680	(2.90)**						
1930s, Huang-Huai-Hai			1.5535	(2.90)**				
1950s, whole sample			0.1195	(0.68)			0.4140	(0.77)
1950s, desert-steppe	0.4193	(1.84)[a]						
1950s, spring wheat	0.0007	(0.01)						
1950s, winter wheat–millet	0.1359	(0.70)						
Agricultural region	*(vs. winter wheat–millet)*				*(vs. north Chang Jiang)*			
Desert-steppe region	-0.1387	(0.75)			-1.4500	(3.19)**		
Spring wheat region	0.2469	(3.23)**			-1.3322	(3.36)		
Huang-Huai-Hai plain					-0.1269	(0.47)**		
South Chang Jiang plain								
Double-cropping rice					-1.3629	(4.83)**		

Table A.2
(continued)

Independent variables	Equation numbers and dependent variables							
	(1) Northwest		(2) The Huang-Huai-Hai plain		(3) Coast		(4) Northeast provinces	
	Coeff.	$|t|$	Coeff.	$|t|$	Coeff.	$|t|$	Coeff.	$|t|$
Densities of human employment in each type of county in 1982								
ln (agricultural density), T	0.0078	(0.23)	0.0570	(1.74)^b	−0.4024	(2.68)**	0.0106	(0.10)
ln (nonagricultural), T	0.0094	(0.29)	−0.0217	(0.64)	0.4693	(3.70)**	0.0256	(0.29)
ln (agricultural density), NL	0.0847	(1.41)	0.1115	(1.60)^b	−0.2781	(1.14)	0.1245	(0.84)
ln (nonagricultural), NL	0.0012	(0.03)	−0.0088	(0.20)	0.6159	(2.31)*	0.0751	(0.58)
No density given					0.2443	(0.54)	−0.2539	(0.47)
Landscape descriptions at profile site (other than tilled fields)								
Grass	−0.0378	(0.44)	0.1743	(1.07)	−0.0845	(0.14)	0.3605	(1.15)
Marsh	0.1232	(0.91)	−0.1244	(1.64)^b	−0.0348	(0.07)	0.8731	(3.14)**
Forest	−0.0853	(0.80)	−0.0757	(0.41)	−0.9321	(1.33)	0.0651	(0.26)
Waste	−0.1066	(0.94)	−0.0094	(0.06)	−0.2401	(0.44)	0.1535	(0.53)
Hill	−0.0068	(0.09)	−0.0732	(0.95)			0.1399	(0.74)
Steep	0.0151	(0.17)	0.1302	(0.63)			−0.1179	(0.43)
Mountain area	−0.2723	(2.84)**	−0.1211	(0.84)			0.0665	(0.30)
No landscape description given	−0.3253	(1.76)^b	−0.1852	(0.57)	−0.6574	(0.77)	−0.4954	(1.31)
Parent material (other than alluvial)								
Weathered crystalline	−0.4855	(4.86)**	−0.9477	(7.05)**			−0.0161	(0.55)

Table A.2
(continued)

Independent variables	Equation numbers and dependent variables							
	(1) Northwest		(2) The Huang-Huai-Hai plain		(3) Coast		(4) Northeast provinces	
	Coeff.	|t|	Coeff.	|t|	Coeff.	|t|	Coeff.	|t|
Weathered clastic	-0.1136	(1.26)	-0.4350	(3.06)**			0.0623	(0.17)
Weathered calcareous	0.0145	(0.12)	-0.0042	(0.03)			0.8230	(1.95)[a]
Diluvial	-0.2476	(3.41)**	-0.3767	(3.08)**			0.2670	(1.08)
Quaternary red clay	-0.0808	(0.49)	0.1628	(0.60)			0.2041	(0.29)
Lacustrine deposits	-0.0620	(0.38)	0.2035	(1.30)			-0.0967	(0.20)
Coastal marine deposits			-0.2484	(1.41)				
Loess	-0.0388	(0.54)	-0.0790	(0.53)			0.2250	(0.82)
Aeolian sands	-0.5252	(2.86)**	-0.0931	(0.19)			0.6801	(1.14)
Soil classifications (other than chao tu fluv-aquic; from Agricultural Atlas)								
Classes 1–9	-1.4023	(10.4)**						
Latosols					-0.0088	(0.02)		
Yellow-brown earths			-0.6166	(3.38)**	-0.3079	(0.87)		
Brown earths			-0.5818	(4.47)**				
Gray forest soils							-0.2027	(0.58)
Cinnamon soils	-0.3058	(2.72)**	-0.2069	(1.81)[a]			0.4073	(1.09)
Loutu, old loessial	-0.0265	(0.14)						
Gray-cinnamon forest	-0.4154	(2.64)**						

Table A.2
(continued)

	Equation numbers and dependent variables							
	(1) Northwest		(2) The Huang-Huai-Hai plain		(3) Coast		(4) Northeast provinces	
Independent variables	Coeff.	|t|	Coeff.	|t|	Coeff.	|t|	Coeff.	|t|
Heilutu, dark loessial	−0.3191	(1.81)[a]						
Chernozems	−1.1362	(7.66)**					0.5105	(1.43)
Chestnut soils	−0.3491	(3.07)**					0.4019	(0.99)
Gray desert soils	−0.0636	(0.50)						
Brown desert soils	0.0992	(0.30)						
Black (*heitu*)	−1.6751	(7.06)**	0.0423	(0.14)			−0.0702	(0.22)
Albic (*baijiangtu*)							−1.0050	(2.85)**
Calcic black			−0.4541	(2.89)**				
Meadow soils	−0.2332	(1.71)[b]	−0.6542	(2.66)**			0.7907	(3.71)**
Warped irrigated	−0.3236	(1.97)[a]						
Bog soils	−0.3987	(2.62)**						
Paddy soils	−0.3961	(2.40)*	−0.2840	(2.11)*	−1.1633	(5.83)**	0.4913	(1.55)
Saline soils	0.8852	(6.40)**	0.4264	(3.59)**	0.2361	(0.97)	1.7633	(6.96)**
Alkali soils			1.1593	(4.86)**			2.2100	(5.77)**
Purple soils	−0.2393	(0.57)						
Limestone soils							−0.4866	(0.89)
Yellow cult. loessial	−0.2826	(1.83)[a]						
Aeolian sandy soils	0.2353	(1.01)	−0.1190	(0.22)			−0.8832	(1.32)

Table A.2
(continued)

| | Equation numbers and dependent variables | | | | | | | |
Independent variables	(1) Northwest		(2) The Huang-Huai-Hai plain		(3) Coast		(4) Northeast provinces	
	Coeff.	$|t|$	Coeff.	$|t|$	Coeff.	$|t|$	Coeff.	$|t|$
Subalpine and alpine	-1.1101	(5.66)**					-1.5328	(2.92)**
Alluvial, fluvisols	-0.1194	(0.80)	-0.0580	(0.19)			0.0374	(0.12)
Rocky	-0.5246	(2.89)**	-0.2652	(1.73)[b]			0.0083	(0.02)
Red clay soils	-0.4606	(2.26)*	-0.8189	(2.40)*			0.4512	(0.77)
Andept, volcanic ash							-0.4103	(0.76)
Miscellaneous soils			-0.1991	(0.85)	0.2184	(0.54)		
Precipitation (meters per year)								
Precipitation	0.26179	(0.40)	0.0359	(0.01)	-5.7914	(3.74)**	-10.2860	(4.24)**
Precipitation squared	-0.8404	(1.27)	-0.6645	(0.34)	1.6608	(3.29)**	5.8656	(3.51)**
Other variables								
Shallow, (cm)$^{-1}$	-0.0129	(0.03)	-0.3127	(0.50)	3.3748	(1.45)	-0.9440	(0.98)
Sand %	0.0171	(1.33)	-0.0085	(0.52)	-0.0533	(1.23)	0.0100	(0.24)
Sand % squared	-0.00015	(1.62)[b]	2E-06	(0.02)	0.0004	(1.43)	0.000062	(0.21)
Clay %	-0.0249	(1.10)	-0.0363	(1.13)	-0.0407	(0.70)	0.0152	(0.28)
Clay % squared	0.00038	(1.21)	0.0005	(1.34)	0.0001	(0.21)	-0.00033	(0.57)
Sand times clay	0.00015	(0.58)	0.0003	(0.81)	0.0010	(1.38)	-0.00025	(0.36)
No texture given	0.0388	(0.08)	-0.8519	(1.30)	-1.2011	(0.68)	0.7237	(0.46)
Source = Thorp map	-0.0800	(0.90)	-0.4616	(2.42)*	0.0777	(0.24)		

Table A.2
(continued)

Independent variables	Equation numbers and dependent variables							
	(1) Northwest		(2) The Huang-Huai-Hai plain		(3) Coast		(4) Northeast provinces	
	Coeff.	$|t|$	Coeff.	$|t|$	Coeff.	$|t|$	Coeff.	$|t|$
Implicit constant	8.0736	(15.6)**	8.1643	(6.32)**	13.1090	(6.51)**	10.1670	(5.61)**
GLS equation constant	0.3381	(1.01)	1.4184	(2.24)*				
Adjusted R-squared	0.9498		0.8787		0.7539		0.6458	
Dependent variable mean	8.4972		7.8614		7.4273		7.0611	
Standard error of estimate	0.5452		0.5275		0.6404		0.6539	
Least-squares type type	GLS		GLS		OLS		OLS	
Number of observations, degrees of freedom	682, 620		492, 442		126, 95		261, 210	

Notes: Statistical significance (two-tailed): **1-percent level, *5-percent, [a]7-percent, [b]10-percent. For GLS equations the dependent variable and its standard error are those of the implicit equation, not the transformed equation in which all variables have been divided by a predicted standard error. T indicates that the profile itself is from a tilled field. NL indicates that no landscape description is given for this profile.

Table A.3
Determinants of alkalinity in north China, 1930s–1980s (full equations)

	Equation numbers			
	(1) Northwest		(2) The Huang-Huai-Hai plain	
Independent variables	Coeff.	|t|	Coeff.	|t|
Time period and region (relative to 1980s, same region; – = positive growth up to 1980s)				
1930s, whole sample			-0.0304	(0.28)
1930s, coast			0.9005	(3.01)**
1930s, desert-steppe	0.2569	(1.03)		
1930s, spring wheat	-0.1419	(1.30)		
1930s, winter wheat–millet	0.2280	(2.13)*		
1950s, whole sample			0.2665	(2.02)*
1950s, desert-steppe	0.1463	(0.71)		
1950s, spring wheat	0.1381	(1.90)[a]		
1950s, winter wheat–millet	0.1146	(0.70)		
Agricultural region	*(vs. winter wheat–millet)*			
Desert-steppe region	-0.0199	(0.11)		
Spring wheat region	0.2402	(2.99)**		
Densities of human employment in that county in 1982				
ln (agricultural density), T	-0.0095	(0.28)	-0.0550	(1.49)
ln (nonagricultural), T	0.0970	(1.52)	0.0053	(0.13)
ln (agricultural density), NL	0.0026	(0.09)	-0.0284	(0.47)

Table A.3
(continued)

Independent variables	Equation numbers			
	(1) Northwest		(2) The Huang-Huai-Hai plain	
	Coeff.	|t|	Coeff.	|t|
ln (nonagricultural), NL	−0.0080	(0.16)	−0.0105	(0.22)
No density given	0.0933	(0.69)		
Landscape descriptions at profile site (other than tilled fields)				
Grass	−0.0403	(0.46)	−0.2585	(1.19)
Marsh	0.0378	(0.28)	−0.1085	(1.20)
Forest	−0.0339	(0.29)	−0.2512	(0.81)
Waste	−0.0927	(0.81)	−0.3080	(1.98)[a]
Hill	0.0326	(0.45)	−0.2867	(2.07)*
Steep	−0.0499	(0.59)	0.7702	(1.58)
Mountain area	−0.0399	(0.37)	−0.2152	(0.51)
No landscape description given	−0.4622	(2.47)*	−0.0221	(0.08)
Parent material (other than alluvial)				
Weathered crystalline	−0.3258	(2.93)**	−0.8065	(2.29)*
Weathered clastic	−0.0577	(0.67)	−3.4973	(0.00)
Weathered calcareous	−0.1037	(0.81)	−0.4388	(1.73)[b]
Diluvial	−0.2157	(2.96)**	−0.4738	(2.02)*
Quarternary red clay	0.0977	(0.68)	0.5221	(1.29)
Lacustrine deposits	−0.0436	(0.29)	−0.0430	(0.23)

Table A.3
(continued)

Independent variables	Equation numbers			
	(1) Northwest		(2) The Huang-Huai-Hai plain	
	Coeff.	|t|	Coeff.	|t|
Coastal marine deposits	-0.0478	(0.67)	-0.4094	(2.02)*
Loess	-0.4876	(2.76)**	0.1629	(0.82)
Aeolian sands			-2.7305	(0.00)
Soil Classifications (from agricultural atlas)				
Classes 1–9	-1.0066	(6.62)**		
Yellow-brown earths			-4.1062	(0.00)
Brown earths			-0.6108	(2.35)*
Cinnamon soils	-0.3602	(3.38)**	-0.3043	(1.77)[b]
Loutu, old loessial	-0.0299	(0.17)		
Gray-cinnamon forest	-0.6904	(4.26)**		
Heilutu, dark loessial	-0.3791	(2.38)*		
Chernozems	-0.9126	(6.29)**		
Chestnut soils	-0.2680	(2.50)*		
Gray desert soils	-0.0131	(0.11)		
Brown desert soils	0.2024	(0.68)		
Black (*heitu*)	-3.1277	(0.01)	-3.4883	(0.00)
Albic (*baijiangtu*)				
Calcic black			-0.1290	(0.68)
Meadow soils	-0.0126	(0.11)	-0.3767	(1.18)

Table A.3
(continued)

Independent variables	Equation numbers			
	(1) Northwest		(2) The Huang-Huai-Hai plain	
	Coeff.	|t|	Coeff.	|t|
Warped irrigated	−0.2658	(1.79)[b]		
Bog soils	−0.3761	(2.73)**		
Paddy soils	−0.4290	(2.20)*	−0.4863	(2.62)**
Saline soils	0.8721	(6.79)**	0.2308	(1.97)[a]
Alkali soils			0.9624	(4.07)**
Purple soils	−0.4805	(1.28)		
Yellow cult. loessial	−0.2671	(1.82)[a]		
Aeolian sandy soils	0.1680	(0.78)	2.6725	(0.00)
Subalpine and alpine	−0.7147	(3.26)**		
Alluvial, fluvisols	−0.2248	(1.58)	−0.2871	(0.83)
Rocky	−0.3422	(2.00)*	−0.0378	(0.12)
Red clay soils	−0.8693	(3.53)**	−4.1129	(0.00)
Miscellaneous soils			−3.5391	(0.00)
Precipitation (meters per year)				
Precipitation	0.1955	(0.28)	−1.5426	(0.50)
Precipitation squared	−0.1597	(0.21)	1.4144	(0.69)
Other variables				
Shallow, $(\text{cm})^{-1}$	−0.0463	(0.91)	−0.6246	(0.77)
Sand %	0.0110	(0.03)	0.0113	(0.45)
Sand % squared	−0.00008	(0.00)	−0.000066169	(0.39)

Table A.3
(continued)

Independent variables	Equation numbers			
	(1) Northwest		(2) The Huang-Huai-Hai plain	
	Coeff.	|t|	Coeff.	|t|
Clay %	-0.0108	(0.05)	0.0036	(0.09)
Clay % squared	0.00018	(0.00)	0.0001	(0.23)
Sand times clay	0.00006	(0.00)	-0.0006	(1.02)
No texture given	0.0866	(0.98)	0.1838	(0.20)
Source = Thorp map	-0.0595	(0.25)	-0.2644	(1.25)
Constant	0.2365	(1.08)	0.6513	(0.46)
Adjusted R-squared	0.5531		0.4752	
Dependent variable mean	0.2278		0.0077	
Standard error of estimate	0.2976		0.2203	
Regression type	tobit		tobit	
Number of observations, degrees of freedom	682, 620		492, 442	

Notes: Dependent variable is Alk = max (pH − 8.0, 0). Statistical significance (two-tailed): **1-percent level, *5-percent, [a]7-percent, [b]10-percent. T indicates that the profile itself is from a tilled field. NL indicates that no landscape description is given for this profile. For tobit equations the dependent-variable mean is the expected value of Alk at mean values of the independent variables, the standard error is the root mean square error, and the R-squared is the squared correlation of observed and expected values. On the use of tobit regessions for dealing with bounded dependent variables (such as Alk, bounded at zero), see Maddala 1983.

Table A.4
Determinants of soil nutrient levels in south China, 1930s–1980s (full equations)

| | Equation numbers and dependent variables | | | | | | | |
| | (1) ln (percentage organic matter) | | (2) ln (percentage nitrogen) | | (3) ln (percentage phosphorus) | | (4) ln (percentage potassium) | |
Independent variables	Coeff.	|t|	Coeff.	|t|	Coeff.	|t|	Coeff.	|t|
Time period and region (relative to 1980s, same region; − = positive growth up to 1980s)								
1930s, all south China	−0.2853	(1.60)	−0.1286	(0.64)	−0.2048	(1.40)	−0.9951	(5.61)**
1950s, all south China					0.0618	(0.48)	−0.3321	(2.44)*
1950s, southwest China	−0.4852	(3.61)**	−0.1528	(1.09)				
1950s, other south China	0.0406	(0.35)	0.1961	(1.79)[b]				
Agricultural region (other than north Chang Jiang plain)								
Desert-steppe region	0.2553	(0.74)	0.1930	(0.61)	0.5164	(1.76)[b]	0.0652	(0.23)
Sichuan basin	−0.0435	(0.57)	−0.0830	(1.16)	−0.0310	(0.36)	−0.2153	(3.37)**
South Chang Jiang plain	0.0554	(0.94)	−0.0024	(0.04)	−0.0650	(0.85)	−0.1529	(3.03)**
Double-cropping rice	0.1165	(1.54)	−0.0457	(0.58)	0.1346	(1.56)	−0.3098	(3.24)**
Southwestern rice	0.4365	(5.14)**	0.3915	(5.26)**	0.2516	(2.95)**	0.5268	(4.83)**
Densities of human employment in that county in 1982								
ln (agricultural density), T	−0.0728	(2.71)**	−0.0560	(2.58)*	0.0054	(0.22)	−0.0240	(1.07)
ln (nonagricultural), T	0.0625	(2.64)**	0.0581	(2.75)**	0.0565	(2.17)*	−0.0102	(0.49)
ln (agricultural density), NL	−0.0854	(1.55)	−0.0773	(1.41)	−0.5415	(0.94)	−0.1172	(2.14)*
ln (nonagricultural), NL	−0.0307	(0.91)	−0.0263	(0.81)	0.0319	(0.85)	−0.0001	(0.00)

Table A.4
(continued)

Independent variables	Equation numbers and dependent variables							
	(1) ln (percentage organic matter)		(2) ln (percentage nitrogen)		(3) ln (percentage phosphorus)		(4) ln (percentage potassium)	
	Coeff.	|t|	Coeff.	|t|	Coeff.	|t|	Coeff.	|t|
Landscape descriptions at profile site (other than tilled fields)								
Grass	-0.1066	(1.08)	-0.1339	(1.39)	-0.0417	(0.40)	-0.1338	(1.31)
Marsh	-0.2917	(1.53)	0.2708	(1.44)	0.3967	(2.20)*	0.0831	(0.37)
Forest	0.0484	(0.59)	-0.0318	(0.39)	-0.0612	(0.69)	-0.1072	(1.27)
Waste	-0.3144	(2.61)*	-0.2814	(2.41)*	-0.1212	(0.97)	-0.1546	(1.26)
Hill	-0.0576	(1.08)	-0.1128	(2.17)*	-0.0131	(0.20)	-0.0366	(0.72)
Steep	-0.0380	(0.49)	-0.0971	(1.25)	0.0324	(0.31)	-0.0605	(0.95)
Mountain area	0.1856	(2.68)**	0.2212	(3.24)**	0.1132	(1.33)	0.0487	(0.76)
No landscape description given	0.3804	(1.92)	0.2460	(1.05)	0.3078	(1.27)	0.6427	(2.77)**
Parent material (other than alluvial)								
Weathered crystalline	0.0685	(0.93)	-0.0448	(0.62)	0.0508	(0.57)	-0.0392	(0.58)
Weathered clastic	-0.0211	(0.31)	0.0087	(0.13)	-0.0826	(1.00)	-0.0842	(1.34)
Weathered calcareous	-0.0604	(0.71)	0.0626	(0.76)	-0.0436	(0.44)	0.1895	(2.30)*
Diluvial	-0.0768	(0.95)	-0.0415	(0.52)	-0.0564	(0.59)	-0.1155	(1.53)
Quaternary red clay	-0.2720	(3.18)**	0.2173	(2.60)**	-0.0631	(0.61)	-0.3782	(4.98)**
Lacustrine deposits	0.2907	(3.66)**	0.2673	(3.31)**	0.0121	(0.11)	-0.0300	(0.42)
Coastal marine deposits	0.0059	(0.08)	-0.0502	(0.65)	0.0012	(0.01)	-0.0127	(0.19)
Loess	-0.3502	(3.75)**	-0.3324	(3.57)**	-0.3202	(2.13)*	-0.2566	(2.87)**
Aeolian sands	0.2762	(0.39)	-0.2806	(0.41)	0.0813	(0.12)	0.8152	(0.97)

Table A.4
(continued)

Independent variables	Equation numbers and dependent variables							
	(1) ln (percentage organic matter)		(2) ln (percentage nitrogen)		(3) ln (percentage phosphorus)		(4) ln (percentage potassium)	
	Coeff.	|t|	Coeff.	|t|	Coeff.	|t|	Coeff.	|t|
Soil classifications (other than chao tu fluvo-aquic; from Agricultural Atlas)								
Latosols	0.0710	(0.51)	−0.1370	(0.99)	−0.6604	(4.20)**	−1.4254	(8.40)**
Laterite red earths	−0.0635	(0.54)	−0.2133	(1.82)[a]	−0.4435	(3.26)**	−0.1886	(1.51)
Red earths	0.1246	(1.32)	0.0257	(0.27)	−0.2428	(2.03)*	−0.1377	(1.57)
Yellow earths	0.4965	(4.91)**	0.3468	(3.44)**	−0.3325	(2.67)**	−0.2599	(2.69)**
Torrid red earths	0.1033	(0.35)	0.5807	(1.56)	−0.3400	(1.06)	0.5128	(1.40)
Yellow-brown earths	0.2343	(2.27)*	0.2033	(1.94)[a]	−0.1793	(1.35)	−0.0167	(0.19)
Brown earths	1.4347	(7.94)**	1.2015	(6.80)**	0.5541	(2.61)**	−0.1513	(1.02)
Forest and podzolic	1.6521	(5.70)**	1.4368	(4.96)**	0.6370	(2.13)*	0.0913	(0.32)
Cinnamon soils	0.7052	(3.00)**	0.6150	(2.73)**	0.2705	(1.25)	−0.1795	(0.90)
Calcic black	0.1310	(0.41)	−0.0068	(0.02)	−0.0606	(0.11)		
Paddy soils	0.4021	(5.83)**	0.3241	(4.54)**	−0.1907	(2.07)*	−0.0484	(0.78)
Saline soils	−0.0130	(0.11)	−0.0767	(0.57)	0.0238	(0.15)	0.1128	(0.99)
Purple soils	−0.1306	(1.23)	0.1431	(1.35)	0.3216	(2.59)*	0.0317	(0.34)
Limestone soils	0.5242	(4.28)**	0.4463	(3.72)**	0.1737	(1.18)	−0.1580	(1.36)
Aeolian soils	−1.4015	(2.04)*	−0.9063	(1.39)	−1.4092	(2.22)*	−1.5561	(1.93)[a]
Subalpine and alpine	1.5974	(4.28)**	1.5664	(4.08)**	0.4776	(1.41)	−0.5816	(1.72)[b]
Alluvial, fluvisols	0.0022	(0.01)	−0.0278	(0.17)	−0.1840	(1.05)	−0.3089	(1.64)[b]
Rocky	−0.0298	(0.25)	−0.0092	(0.08)	−0.3734	(2.52)*	0.2043	(1.60)
Red clay	0.4074	(1.91)[a]	0.2926	(1.35)	−0.2494	(0.89)	−0.1502	(0.72)
Andept, volcanic ash	0.1342	(0.65)	0.1738	(0.84)	0.6452	(2.58)*	−0.3460	(1.93)[a]

Table A.4
(continued)

Independent variables	Equation numbers and dependent variables							
	(1) ln (percentage organic matter)		(2) ln (percentage nitrogen)		(3) ln (percentage phosphorus)		(4) ln (percentage potassium)	
	Coeff.	|t|	Coeff.	|t|	Coeff.	|t|	Coeff.	|t|
Precipitation (meters per year)								
Precipitation	0.0430	(0.10)	0.3149	(0.73)	0.1117	(0.21)	−0.2237	(0.55)
Precipitation squared	0.1098	(0.76)	−0.0179	(0.12)	−0.0635	(0.34)	0.1027	(0.74)
Other variables								
Shallow, (cm)$^{-1}$	0.9445	(2.81)**	0.9380	(2.80)**	0.1955	(0.48)	0.1674	(0.51)
Sand %	0.0133	(1.10)	0.0191	(1.55)	0.0138	(0.90)	0.0037	(0.31)
Sand % squared	−0.00015	(1.76)[b]	−0.00019	(2.26)*	−0.00014972	(1.43)	5.5116E-06	(0.06)
Clay %	0.0358	(2.19)*	0.0405	(2.42)*	0.0039	(0.19)	0.0239	(1.48)
Clay % squared	−0.00045	(2.82)**	−0.00047	(2.85)**	0.00007	(0.34)	−0.00028864	(1.82)[a]
Sand times clay	−0.00020	(0.97)	−0.00028	(1.31)	−0.00016472	(0.65)	−0.0001	(0.46)
No texture given	0.6133	(1.34)	0.7398	(1.60)	0.2199	(0.37)	0.3174	(0.72)
Source = Guangdong 1980s					−0.9917	(4.78)**		
Source = Hunan 1980s					−0.1889	(0.73)		
Source = Guangxi 1980s							−0.2976	(3.10)**
Source = Fujian 1980s							0.3187	(3.00)**
Source = Sichuan 1980s							0.1054	(1.15)
Source = Hubei 1980s							−0.0801	(1.42)
Implicit constant	−0.3991	(0.74)	−3.5653	(6.17)**	3.3385	(4.69)**	0.7082	(1.38)
GLS equation constant	0.0661	(0.18)	0.2593	(0.56)	0.0463	(0.18)	−0.4216	(1.76)[b]

Table A.4
(continued)

	Equation numbers and dependent variables			
	(1) ln (percentage organic matter)	(2) ln (percentage nitrogen)	(3) ln (percentage phosphorus)	(4) ln (percentage potassium)
Adjusted R-squared	0.4057	0.4719	0.5332	0.4948
Dependent variable mean	0.8997	-2.1106	-3.2203	0.4521
Standard error of estimate	0.5578	0.5350	0.6666	0.5594
Least-squares type	GLS	GLS	GLS	GLS
Number of observations, degrees of freedom	1,223 1,163	1,194 1,134	1,182 1,121	1,042 980

Notes: Statistical significance (two-tailed): ** 1-percent level, * 5-percent, [a]7-percent, [b]10-percent. For GLS equations the dependent variable and its standard error are those of the implicit equation, not the transformed equation in which all variables have been divided by a predicted standard error. T indicates that the profile itself is from a tilled field. NL indicates that no landscape description is given for this profile.

Table A.5
Determinants of pH levels and acidity in south China, 1930s–1980s (full equations)

Independent variables	Equation numbers and dependent variables							
	(1)–(2): Southwest, south and Sichuan				(3)–(4): East (Chang Jiang plain)			
	(1) pH level		(2) Acidity		(3) pH level		(4) Acidity	
	Coeff.	|t|	Coeff.	|t|	Coeff.	|t|	Coeff.	|t|
Time period and region (relative to 1980s, same region; – = positive growth up to 1980s)								
1930s, coast					−0.0612	(0.30)	1.1466	(0.00)
1930s, Sichuan basin	0.1142	(0.63)	−0.3024	(1.49)				
1930s, north Chang Jiang					0.0749	(0.53)	−0.0878	(0.52)
1930s, south Chang Jiang					−0.3613	(2.79)**	0.2060	(1.70)[b]
1930s, double-crop rice	−0.7133	(3.77)**	0.9614	(6.14)**				
1930s, southwestern rice	−0.5079	(3.55)**	0.5587	(4.06)**				
1950s, Sichuan basin	−1.3319	(1.89)[a]	1.4910	(2.22)*				
1950s, north Chang Jiang					0.2718	(1.12)	−0.6590	(2.46)*
1950s, south Chang Jiang					0.6017	(2.55)*	−0.5743	(2.81)**
1950s, double-crop rice	0.1061	(0.38)	−0.0545	(0.20)				
1950s, southwestern rice	0.7187	(3.16)**	−0.0911	(2.94)**				
Agricultural region	*(other than Sichuan basin)*				*(other than north Chang Jiang plain)*			
Desert-steppe region	0.2241	(0.67)	0.0401	(0.10)				
South Chang Jiang plain					−0.4085	(4.13)**	0.4032	(4.55)**
Double-cropping rice	−0.2113	(1.13)	0.0676	(0.34)				
Southwestern rice	−0.2182	(1.38)	0.1790	(1.03)				

Table A.5
(continued)

Independent variables	Equation numbers and dependent variables							
	(1)–(2): Southwest, south and Sichuan				(3)–(4): East (Chang Jiang plain)			
	(1) pH level		(2) Acidity		(3) pH level		(4) Acidity	
	Coeff.	|t|	Coeff.	|t|	Coeff.	|t|	Coeff.	|t|
Densities of human employment in that county in 1982								
ln (agricultural density), T	0.0745	(1.54)	−0.0089	(0.19)	−0.0695	(1.80)[a]	0.0266	(0.70)
ln (nonagricultural), T	−0.0018	(0.04)	−0.0861	(1.79)[b]	0.1034	(2.32)*	−0.0821	(1.98)[a]
ln (agricultural density), NL	−0.0112	(0.13)	0.0316	(0.36)	0.1839	(1.26)	−0.1505	(1.08)
ln (nonagricultural), NL	−0.0076	(0.15)	0.0152	(0.31)	−0.0666	(1.12)	0.0502	(0.89)
Landscape descriptions at profile site (other than tilled fields)								
Grass	0.1401	(0.80)	−0.1475	(0.85)	−0.2598	(1.67)[b]	0.4266	(2.49)*
Marsh	−0.1465	(0.62)	0.2082	(0.87)	−0.9022	(1.89)[a]	0.8593	(1.50)
Forest	−0.2364	(1.45)	0.2211	(1.41)	−0.5607	(4.38)**	0.5039	(3.79)**
Waste	0.1952	(1.04)	−0.1304	(0.72)	−0.2429	(1.57)	0.2719	(1.36)
Hilly	0.1149	(0.97)	−0.1028	(0.89)	−0.0210	(0.23)	−0.0953	(1.21)
Steep	−0.1420	(0.61)	0.1467	(0.68)	−0.0423	(0.32)	−0.1163	(1.01)
Mountain area	0.1530	(0.83)	−0.2975	(1.61)[b]	−0.0007	(0.01)	−0.1210	(1.17)
No landscape description given	0.1540	(0.51)	−0.3576	(1.08)	−1.0084	(1.33)	0.8402	(1.16)
Parent material (other than alluvial)								
Weathered crystalline rock	−0.5463	(3.19)**	0.5225	(3.16)**	−0.2920	(2.21)*	0.1721	(1.53)
Weathered clastic	0.3361	(2.35)*	0.3328	(2.32)*	−0.2373	(1.99)[a]	0.1934	(1.87)[a]
Weathered calcareous	0.2478	(1.54)	−0.0634	(0.39)	−0.0266	(0.19)	−0.0297	(0.21)

Table A.5
(continued)

	Equation numbers and dependent variables							
	(1)–(2): Southwest, south and Sichuan				(3)–(4): East (Chang Jiang plain)			
	(1) pH level		(2) Acidity		(3) pH level		(4) Acidity	
Independent variables	Coeff.	|t|	Coeff.	|t|	Coeff.	|t|	Coeff.	|t|
Diluvial	−0.3637	(2.54)*	0.3143	(2.20)*	0.2133	(1.52)	−0.1473	(1.16)
Quaternary red clay	−0.3090	(1.42)	0.1506	(0.72)	−0.3995	(2.59)*	0.2672	(2.12)*
Lacustrine deposits	−0.4250	(1.26)	0.3037	(0.93)	−0.0104	(0.07)	−0.0646	(0.46)
Coastal marine depos.	−0.8423	(3.23)**	0.7471	(3.00)**	0.2773	(1.82)[a]	−0.4714	(2.80)**
Loess	0.4536	(1.45)	−5.5505	(0.00)	−0.2308	(1.63)[b]	0.1276	(0.91)
Soil classifications (other than chao tu fluvo-aquic; from Agricultural Atlas)								
Latosols	0.1890	(0.62)	−0.0500	(0.17)				
Laterite red earths	−0.1272	(0.46)	0.2252	(0.82)	−0.81934	4.2478**	0.4406	(2.30)*
Red earths	−0.2372	(0.87)	0.3731	(1.39)				
Yellow earths	−0.5315	(1.95)[a]	0.7257	(2.69)**	0.66346	2.9796**	0.4247	(2.03)*
Torrid red earths	0.2595	(0.48)	0.0013	(0.00)				
Yellow-brown earths	0.1557	(0.44)	−0.0598	(0.16)	−0.6774	(3.30)**	0.3608	(1.71)[b]
Brown earths	−0.8957	(2.27)*	1.0744	(2.84)**	−0.7997	2.163*	0.1096	(0.25)
Forest and podzolic	0.1599	(0.34)	−0.1165	(0.22)				
Cinnamon soils	0.8518	(2.18)*	−4.6784	(0.00)				
Four dark-soil classes	−0.5030	(0.80)	1.1684	(1.65)[b]				
Calcic black					1.0878	1.8434[a]	−2.4720	(0.00)
Meadow soils	−0.2979	(0.65)	1.0234	(2.29)*	−1.2174	(3.94)**	0.9334	(3.38)**

Table A.5
(continued)

Independent variables	Equation numbers and dependent variables							
	(1)–(2): Southwest, south and Sichuan				(3)–(4): East (Chang Jiang plain)			
	(1) pH level		(2) Acidity		(3) pH level		(4) Acidity	
	Coeff.	$|t|$	Coeff.	$|t|$	Coeff.	$|t|$	Coeff.	$|t|$
Bog soils	−0.5976	(1.48)	0.8709	(2.23)*	−0.8030	(5.43)**	0.5460	(3.40)**
Paddy soils	0.2025	(0.81)	0.0009	(0.00)	0.9932	(5.52)**	3.6670	(0.00)
Saline soils	1.9823	(4.57)**	−1.2064	(2.58)*	0.4027	(1.92)[a]	−0.2382	(1.16)
Purple soils	0.7314	(2.55)*	−0.3189	(1.11)	0.8184	(3.17)**	−4.5783	(0.00)
Limestone soils	0.9526	(3.25)**	−1.0049	(3.18)**				
Aeolian soils	0.9535	(2.25)*	−0.6825	(1.45)				
Subalpine and alpine	−0.2805	(0.76)	0.7315	(2.01)*				
Alluvial, fluvisols	0.4118	(1.27)	0.0676	(0.21)	−0.8202	(1.22)	0.3546	(0.59)
Rocky	0.1869	(0.56)	−0.0503	(0.15)	−0.5338	(1.97)[a]	0.0930	(0.40)
Red clays	0.2964	(0.40)	0.7302	(1.04)	0.9469	(2.81)**	0.5600	(1.86)[a]
Andept, volcanic ash	0.9589	(1.59)	−0.7811	(1.28)	−0.1403	(0.38)	0.1346	(0.33)
Precipitation (meters per year)								
Precipitation	−1.8810	(2.30)*	2.1088	(2.42)*	−2.3254	(3.01)**	2.3246	(3.34)**
Precipitation squared	0.5157	(1.77)[b]	0.5217	(1.72)[b]	0.5200	(2.11)*	−0.5846	(2.79)**
Other variables								
Shallow, $(cm)^{-1}$	−0.1651	(0.27)	−0.3409	(0.58)	−0.7663	(1.33)	0.2784	(0.52)
Sand %	−0.0157	(0.79)	0.0218	(1.17)	−0.3530	(1.53)	0.0314	(1.44)
Sand % squared	0.00010	(0.64)	−0.0001	(0.93)	0.0003	(1.44)	−0.0003	(1.81)[a]
Clay %	−0.0213	(0.81)	0.0373	(1.49)	−0.0365	(1.31)	0.0123	(0.39)

Table A.5
(continued)

Independent variables	Equation numbers and dependent variables							
	(1)–(2): Southwest, south and Sichuan				(3)–(4): East (Chang Jiang plain)			
	(1) pH level		(2) Acidity		(3) pH level		(4) Acidity	
	Coeff.	|t|	Coeff.	|t|	Coeff.	|t|	Coeff.	|t|
Clay % squared	0.00012	(0.44)	−0.0003	(1.16)	0.0002	(0.73)	0.0001	(0.18)
Sand times clay	0.00024	(0.75)	−0.0004	(1.16)	0.0007	(1.79)[b]	−0.0005	(1.17)
No texture given	−0.7696	(1.09)	1.0390	(1.57)	−1.2135	(1.62)[b]	0.6562	(0.82)
Source = Thorp map	−0.5611	(3.05)**			0.2130	(1.53)	−0.3658	(1.83)[a]
Implicit constant	8.2910	(8.81)**	−2.8313	(3.02)**	10.3240	(10.4)**	−3.2523	(3.43)**
GLS equation constant					0.3216	(0.64)		
Adjusted R-squared	0.3854		0.3697		0.9502		0.5449	
Dependent variable mean	5.8717		0.3481		6.8333		0.0695	
Standard error of estimate	0.9305		0.5284		0.7029		0.3401	
Regression type	OLS		tobit		GLS		tobit	
Number of observations, degrees of freedom	860, 798		860, 798		727, 675		727, 675	

Notes: Statistical significance (two-tailed): **—1-percent level, *—5-percent, [a]—7-percent, [b]—10-percent. T indicates that the profile itself is from a tilled field. NL indicates that no landscape description is given for this profile. For GLS equations the dependent variable and its standard error are those of the implicit equation, not the transformed equation in which all variables have been divided by a predicted standard error. For tobit equations, the dependent variable is *Acidity* = max (6 − pH, 0). Its standard error is the root mean square error, and the *R*-squared is the squared correlation between observed and expected values. On the tobit regession approach to dealing with bounded dependent variables (e.g., bounded by zero, as here), see Maddala 1983.

Appendix B: Soil Chemistry Averages by Province for China, 1981–1986

Table B.1
Province average characteristics for tilled soils, China, 1981–1986 (calculated from 1980s sample of soil profiles)

	Organic matter (%)	Total N (%)	Total P (%)	Total K (%)	pH (pH)	Acidity (pH < 6)	Alkalinity (pH > 8)
Northern China							
Heilongjiang	1.82	0.161	0.057	1.616	7.04	0.12	0.31
Jilin	2.83	0.150	0.063	2.056	7.31	0.03	0.05
Liaoning	1.52	0.092	0.030	1.716	7.10	0.02	0.04
Hebei	2.11	0.117	0.059	1.894	7.45	0.00	0.00
Nei Mongol	1.58	0.089	0.037	2.015	8.13	0.01	0.29
Ningxia	1.62	0.093	0.071	1.680	8.18	0.00	0.20
Gansu	1.33	0.094	0.076	1.725	8.13	0.00	0.24
Shaanxi	1.14	0.071	0.050	1.720	7.43	0.03	0.15
Shanxi	0.93	0.050	0.060	1.877	8.13	0.00	0.17
Henan	1.09	0.075	0.051	1.774	7.80	0.00	0.21
Shandong	1.02	0.067	0.045	1.952	7.31	0.03	0.02
Central and southern China							
Jiangsu	2.24	0.128	0.045	1.915	7.22	0.00	0.03
Shanghai	2.52	0.144	0.074	2.233	7.49	0.00	0.02
Zhejiang	2.75	0.150	0.040	1.911	6.02	0.23	0.02
Anhui	1.69	0.104	0.050	1.745	6.30	0.40	0.06
Hubei	1.69	0.099	0.042	1.560	6.38	0.19	0.02
Sichuan	2.47	0.142	0.064	1.683	6.90	0.13	0.03
Guizhou	3.14	0.177	0.072	1.581	6.61	0.27	0.01
Hunan	1.42	0.103	0.038	1.687	6.75	0.19	0.13

Table B.1
(continued)

	Organic matter (%)	Total N (%)	Total P (%)	Total K (%)	pH (pH)	Acidity (pH < 6)	Alkalinity (pH > 8)
Central and southern China							
Jiangxi	2.44	0.127	0.049	1.819	5.61	0.62	0.01
Fujian	2.15	0.101	0.055	1.665	5.50	0.66	0.00
Guangdong	1.96	0.100	0.039	0.950	6.00	0.41	0.04
Guangxi	2.36	0.128	0.051	1.031	6.05	0.43	0.02
Yunnan	3.59	0.171	0.065	1.182	5.93	0.43	0.00

Notes: Averages are based on 761 observations that had good 1985 county data on agriculture. Averages are noisy averages based on observed values, not regression-predicted values. Calculations have adjusted ACID, lnOM, lnN, and (south only) lnK for standardization to a profile running from the surface to 10 cm. Figures for lnP were not so adjusted, because its whole regression equation was insignificant. Averages for OM, N, P, and K are antilogs of log averages, which tend to be somewhat lower than averages of the original values.

Appendix C: Fuller Simultaneous-Equation Estimates of China's Soil-Agriculture System in the 1980s

This appendix presents fuller regression equations for the simultaneous soil-agriculture system than the summaries in chapter 5. Here one finds, among other things, some interactions among soil characteristic terms among inputs in the translog part of the production function. For the very fullest form of the regressions, see the computer printouts in Lindert 1996, appendix F.

Table C.1
Determinants of agricultural yields in China, 1985

North China

Equation numbers and dependent variables

Independent variables	(1) ln (GVAO per cultivated area)		(2) ln (grain per total cultivated area)		(3) ln (main crops per total cultivated area)		(4) ln (grain yield per grain-cultivated area)		(5) ln (main-crop yield per area cultivated in main crops)	
	Coeff.	$|t|$	Coeff.	$|t|$	Coeff.	$|t|$	Coeff.	$|t|$	Coeff.	$|t|$
ln (% organic matter)	−0.425	(1.91)	−0.096	(0.51)	−0.231	(1.00)	−0.096	(0.22)	−0.158	(0.36)
ln (total nitrogen)	−0.0003	(0.00)	0.099	(0.70)	−0.058	(0.35)	−0.065	(0.21)	−0.252	(0.78)
ln (total phosphorus)	0.029	(0.29)	0.010	(0.11)	−0.045	(0.45)	0.141	(0.75)	0.074	(0.38)
ln (P) * ln (% organic matter)	−0.124	(1.77)	0.026	(0.44)	−0.031	(0.42)	−0.043	(0.32)	−0.076	(0.54)
ln (P) * (*Alk + Acid*)	−0.260	(0.83)	−0.296	(1.11)	−0.645	(1.99)	−0.472	(0.77)	−0.701	(1.12)
ln (P) * (% clay)	−0.00005	(0.01)	0.000	(0.09)	0.005	(0.95)	0.0045	(0.44)	0.0087	(0.83)
ln (total potassium)	−0.023	(0.19)	0.091	(0.91)	0.215	(1.74)	−0.443	(1.91)	−0.263	(1.11)
Soil alkalinity	−1.051	(1.08)	−1.041	(1.25)	−2.326	(2.30)	−1.412	(0.74)	−2.258	(1.16)
Nonland inputs per cultivated area, translog										
ln (laborers/cultivated area)	−0.732	(0.66)	−2.436	(2.55)	−1.874	(1.62)	−3.940	(1.81)	−3.371	(1.52)
ln (laborers/cultivated area), squared	0.066	(0.85)	0.206	(3.08)	0.146	(1.81)	0.288	(1.90)	0.242	(1.56)
ln (multiple-cropping index)	1.215	(2.55)	−0.229	(0.52)	0.883	(1.52)	−0.032	(0.03)	0.733	(0.77)
ln (multiple-cropping index), squared	−1.273	(1.86)	0.371	(0.64)	−0.885	(1.18)	−0.898	(0.67)	−2.122	(1.56)

Table C.1
(continued)

	Equation numbers and dependent variables									
	(1) ln (GVAO per cultivated area)		(2) ln (grain per total cultivated area)		(3) ln (main crops per total cultivated area)		(4) ln (grain yield per grain-cultivated area)		(5) ln (main-crop yield per area cultivated in main crops)	
Independent variables	Coeff.	|t|	Coeff.	|t|	Coeff.	|t|	Coeff.	|t|	Coeff.	|t|
ln (fertilizer/cultivated area)	0.917	(1.30)	2.844	(4.48)	2.504	(3.31)	4.002	(2.90)	3.425	(2.43)
ln (fertilizer/cultivated area), squared	0.043	(1.50)	0.063	(2.49)	0.063	(1.99)	0.063	(1.12)	0.071	(1.24)
ln (irrigated area/cultivated area)	0.541	(1.05)	0.638	(1.41)	-0.772	(1.45)	2.071	(2.06)	1.753	(1.71)
ln (irrigated area/cultivated area), squared	0.003	(0.11)	-0.037	(1.45)	-0.037	(1.27)	-0.012	(0.22)	-0.025	(0.43)
ln (electric power use/cultivated area)	-0.294	(0.53)	1.154	(2.37)	0.431	(0.70)	0.528	(0.49)	0.284	(0.26)
ln (electric power use/cultivated area), squared	0.037	(1.63)	-0.039	(1.97)	-0.022	(0.87)	-0.100	(2.26)	-0.100	(2.21)
ln (lab./cult.) * ln (fert./cult.)	-0.052	(0.54)	-0.293	(3.45)	-0.234	(2.30)	-0.442	(2.38)	-0.342	(1.80)
ln (lab./cult.) * ln (irrig./cult.)	-0.035	(0.52)	0.049	(0.85)	0.091	(1.33)	-0.358	(2.78)	-0.314	(2.38)
ln (lab./cult.) * ln (elec./cult.)	0.008	(0.11)	-0.109	(1.75)	-0.030	(0.39)	0.055	(0.39)	0.082	(0.56)
ln (fert./cult.) * ln (irrig./cult.)	0.070	(1.51)	0.076	(1.79)	0.104	(2.09)	0.1380	(1.53)	0.155	(1.68)
ln (fert./cult.) * ln (elec./cult.)	-0.050	(1.16)	-0.014	(0.35)	-0.060	(1.34)	-0.0357	(0.42)	-0.098	(1.13)
ln (irrig./cult.) * ln (elec./cult.)	-0.056	(1.37)	0.076	(2.02)	0.022	(0.48)	0.206	(2.60)	0.202	(2.48)
ln (crop share) * ln (lab./cult.)	0.127	(1.08)	-0.298	(0.72)						

Table C.1
(continued)

	Equation numbers and dependent variables									
	(1) ln (GVAO per cultivated area)		(2) ln (grain per total cultivated area)		(3) ln (main crops per total cultivated area)		(4) ln (grain yield per grain-cultivated area)		(5) ln (main-crop yield per area cultivated in main crops)	
Independent variables	Coeff.	|t|	Coeff.	|t|	Coeff.	|t|	Coeff.	|t|	Coeff.	|t|
ln (crop share) * ln (multiple-cropping index)	−0.277	(0.26)	3.396	(0.96)						
ln (crop share) * ln (fert./cult.)	0.395	(1.68)	0.993	(1.32)						
ln (crop share) * ln (irrig./cult.)	−0.331	(1.68)	−1.221	(1.87)						
ln (crop share) * ln (elec./cult.)	−0.127	(0.65)	0.345	(0.63)						
Number of other variables:	11		11		11		11		11	
Adjusted R-squared	0.836		0.880		0.829		0.678		0.648	
Dependent variable mean	8.184		7.603		7.972		3.354		3.551	
Standard error of estimate	0.247		0.207		0.251		0.482		0.493	
Regression type	2SLS		2SLS		2SLS		2SLS		2SLS	
ln (% organic matter)	1.371	(5.37)	0.482	(2.10)	0.600	(2.70)	0.347	(0.44)	0.459	(0.60)
ln (% organic matter) * (% clay)	−0.001	(0.51)	0.000	(0.01)	0.001	(0.34)	−0.001	(0.12)	0.000	(0.03)
ln (total nitrogen)	−0.753	(4.37)	−0.227	(1.44)	−0.321	(2.14)	0.130	(0.25)	0.030	(0.06)
ln (total phosphorus)	−0.149	(2.37)	−0.138	(2.46)	−0.149	(2.72)	−0.360	(1.86)	−0.345	(1.82)
ln (P) * ln (% organic matter)	0.173	(3.36)	0.094	(2.07)	0.137	(3.07)	0.064	(0.40)	0.090	(0.58)
ln (P) * (Alk + Acid)	−0.190	(2.33)	−0.054	(0.74)	−0.068	(0.97)	−0.103	(0.41)	−0.117	(0.48)
ln (P) * (% clay)	0.0005	(0.27)	0.000	(0.20)	0.000	(0.06)	0.006	(1.19)	0.006	(1.07)

Table C.1
(continued)

South China

	Equation numbers and dependent variables									
	(1) ln (GVAO per cultivated area)		(2) ln (grain per total cultivated area)		(3) ln (main crops per total cultivated area)		(4) ln (grain yield per grain-cultivated area)		(5) ln (main-crop yield per area cultivated in main crops)	
Independent variables	Coeff.	|t|	Coeff.	|t|	Coeff.	|t|	Coeff.	|t|	Coeff.	|t|
ln (total potassium)	0.088	(1.55)	0.171	(3.26)	0.149	(3.00)	0.017	(0.10)	0.007	(0.04)
ln (total K) * (Acid)	0.0066	(0.07)	0.024	(0.27)	-0.008	(0.09)	0.320	(1.07)	0.306	(1.04)
Soil acidity	-0.795	(2.92)	-0.215	(0.87)	-0.391	(1.64)	-0.142	(0.17)	-0.279	(0.34)
Nonland inputs per cultivated area, translog										
ln (laborers/cultivated area)	-2.299	(2.13)	-1.427	(1.50)	-1.069	(1.14)	-12.667	(3.84)	-12.110	(3.74)
ln (laborers/cultivated area), squared	0.171	(2.81)	0.133	(2.47)	0.111	(2.08)	0.524	(2.81)	0.493	(2.69)
ln (predicted multiple-cropping index)	-0.046	(0.16)	0.415	(1.49)	0.542	(1.99)	-0.283	(0.32)	-0.161	(0.18)
ln (predicted multiple-cropping index), squared	0.136	(0.62)	-0.026	(0.13)	-0.134	(0.69)	0.407	(0.61)	0.318	(0.49)
ln (predicted fertilizer/cultivated area)	0.805	(1.53)	0.622	(1.34)	0.218	(0.48)	9.696	(6.02)	9.398	(5.94)
ln (predicted fertilizer/cultivated area), squared	0.053	(2.41)	0.043	(2.08)	0.019	(1.00)	0.273	(4.09)	0.259	(3.95)
ln (irrigated area/cultivated area)	-0.208	(0.26)	0.682	(0.97)	0.506	(0.74)	-8.749	(3.55)	-8.716	(3.61)

Table C.1
(continued)

	Equation numbers and dependent variables									
	(1) ln (GVAO per cultivated area)		(2) ln (grain per total cultivated area)		(3) ln (main crops per total cultivated area)		(4) ln (grain yield per grain-cultivated area)		(5) ln (main-crop yield per area cultivated in main crops)	
Independent variables	Coeff.	|t|	Coeff.	|t|	Coeff.	|t|	Coeff.	|t|	Coeff.	|t|
ln (irrigated area/cultivated area), squared	0.112	(3.88)	0.117	(4.61)	0.118	(4.76)	0.107	(1.21)	0.103	(1.19)
ln (electric power use/cultivated area)	−0.023	(0.07)	−0.327	(1.17)	−0.018	(0.07)	−0.482	(0.49)	−0.317	(0.33)
ln (electric power use/cultivated area), squared	0.020	(1.97)	0.005	(0.56)	−0.001	(0.12)	0.045	(1.48)	0.045	(1.51)
ln (lab./cult.) * ln (fert./cult.)	−0.021	(0.34)	0.002	(0.03)	0.043	(0.82)	−1.011	(5.36)	−0.983	(5.30)
ln (lab./cult.) * ln (irrig./cult.)	−0.048	(0.47)	−0.169	(1.90)	−0.143	(1.64)	1.085	(3.48)	1.086	(3.55)
ln (lab./cult.) * ln (elec./cult.)	−0.008	(0.21)	−0.002	(0.06)	−0.039	(1.20)	0.170	(1.47)	0.149	(1.32)
ln (fert./cult.) * ln (irrig./cult.)	−0.110	(2.58)	−0.088	(2.23)	−0.086	(2.31)	−0.343	2.6149	−0.334	2.5931
ln (fert./cult.) * ln (elec./cult.)	−0.023	(0.86)	−0.055	(2.27)	−0.042	(1.70)	−0.039	0.4657	−0.024	0.3001
ln (irrig./cult.) * ln (elec./cult.)	−0.001	(0.02)	0.090	(2.85)	0.089	(2.89)	−0.269	2.5053	−0.270	2.5569
ln (crop share) * ln (lab./cult.)	−0.141	(1.37)	−0.269	(2.30)						
ln (crop share) * ln (multiple-cropping index)	−0.704	(1.56)	−1.155	(2.24)						
ln (crop share) * ln (fert./cult.)	0.086	(0.56)	−0.112	(0.63)						
ln (crop share) * ln (irrig./cult.)	0.426	(1.82)	0.785	(2.92)						
ln (crop share) * ln (elec./cult.)	0.122	(1.33)	0.072	(0.63)						

Table C.1
(continued)

	Equation numbers and dependent variables									
	(1) ln (GVAO per cultivated area)		(2) ln (grain per total cultivated area)		(3) ln (main crops per total cultivated area)		(4) ln (grain yield per grain-cultivated area)		(5) ln (main-crop yield per area cultivated in main crops)	
Independent variables	Coeff.	|t|	Coeff.	|t|	Coeff.	|t|	Coeff.	|t|	Coeff.	|t|
Number of other independent variables	17		17		17		17		17	
Adjusted R-squared	0.823		0.881		0.912		0.429		0.430	
Dependent variable mean	8.696		8.183		8.354		4.775		4.837	
Standard error of estimate	0.174		0.152		0.148		0.534		0.525	
Regression type	2SLS		2SLS		2SLS		2SLS		2SLS	

Notes: The samples: North—266 soil profiles from 164 northern counties; South—495 soil profiles from 287 southern counties. Lindert 1996 gives full regression printouts for the case with instrumented (endogenous) multiple-cropping and fertilizer.

Dependent variables

ln (grain per total cultivated area) = natural log of metric tons of harvested grain per 10,000 mu of area cultivated in that county in 1985.

ln (main crops per total cultivated area) = natural log of tons of grain and cotton and oils per cultivated 10,000 mu, expressed in tons of grain equivalent. Cotton and oils are converted into grain equivalents using the following 1986 average state prices (yuan/ton): grains = 465.76, cotton = 3,576.8, and vegetable oils = 2,628.54 (from Kueh and Ash 1993, pp. 62, 132).

Soil nutrient levels refer to total levels (percentages), not to available levels.

Organic matter percentage has been censored to exclude a small number of available histosol profiles for which the organic matter share would be over 20 percent.

Table C.1
(continued)

Levels of organic matter, nitrogen, and potassium have been adjusted to their estimated levels at a depth of 10 cm from the surface, using the coefficients on depth from the fuller version of equations (6), (7), and (9).

Independent variables

Each variable under an "endogenous" heading takes on values predicted from a first-stage regression making it a function of all exogenous variables used in any of the structural equations (1)–(10).

ln (laborers/cultivated area) = natural log of agricultural laborers in 1982 per 10,000 mu cultivated in 1985.

ln (fertilizer/cultivated area) = natural log of fertilizer tons per 10,000 mu of area cultivated in that county in 1985. Again, 1985 county data.

ln (multiple-cropping index) = natural log of the ratio of sown area to cultivated area in that county in 1985.

ln (irrigated area/cultivated area) = natural log of the share of cultivated area that was irrigated in 1985.

ln (electric power use/cultivated area) = natural log of electricity use per cultivated area, in kilowatt-hours per mu.

"Crop share" refers to the share of total sown area planted in these crops (grains in eq. (1), grains plus cotton and oils in (2)).

Acidity and alkalinity are in the A horizon.

For equations (1)–(3), the additional exogenous variables are four climate variables, five texture variables, soil acidity (north or alkalinity (south), six *Yunnan* variables for the south, and [for (1) and (2) only] the share of these crops in total sown area.

For equations (4) and (5), the additional variables are the same four climate variables, soil acidity, irrigated share, crop-regional variables (three in North, four in South), six *Yunnan* variables (south only—see below), and twelve variables that should influence the government's fertilizer allocation policy across provinces and across counties. These last 12 variables measure crop mix, the share of crop output purchased by the state, the ratio of collective output to the gross value of agricultural output, and the nonagricultural share of the local labor force in 1982. Each of these was posited to be an influence on the lobbying power of counties in getting fertilizer allocations. See Lindert 1996 for the coefficients and tests on these variables.

Equations (6)–(10) include forty-four other exogenous variables for the north and forty-three for the south. Four climate variables and two human-employment density variables refer to the whole county. The others refer to the soil sample site: two terrain conditions, nine parent materials, soil classes (twenty in north, fifteen in south), six texture parameters, the depth of the midpoint of the A-horizon, and four *Yunnan* variables (south only—see below).

The four climate variables were mean annual precipitation (meters), precipitation squared, mean annual temperature (°C), and temperature squared.

Table C.1
(continued)

The crop regions named in Lindert 1996 are those mapped in Buck 1937; Lindert, Lu, and Wu 1996a, 1996b; and Lindert 1996. The northeast provinces of Liaoning and Jilin are counted as part of the "spring wheat" region. All samples omit Heilongjiang, the desert far west (Qinghai, Tibet, Xinjiang), and the islands.

The two binary terrain variables indicated whether the (tilled) soil sample site was marshy or sloped.

The density of human employment in that county in 1982 takes two forms:

(a) agricultural density—the natural log of the number of persons employed in agriculture per square kilometer.

(b) nonagricultural density—the natural log of the number of persons employed in nonagricultural pursuits per square kilometer.

The nine parent material categories (compared to alluvial deposits) are weathered crystalline rock, weathered clastic sedimentary rock, weathered calcareous rock, diluvial deposits, quaternary red clay, lacustrine deposits, coastal marine deposits, loesses, and aeolian sands. These were taken from the 1986 *Soil Atlas* if they were not given in the soil profile data source.

The soil classes (twenty in the north, fifteen in the south) are large group classes taken from the *Agricultural Atlas* of China. All classes are compared with the omitted *chao tu fluvo-aquic*, a common alluvial soil class.

The texture of the profile's A horizon is summarized by the quadratic combination of five percentage shares: sand (and gravel), sand squared, clay, clay squared, and sand times clay. The silt percentage and its square are thus implicit, since silt = 100 percent − sand − clay. The sixth texture variable is a binary for texture not given.

Depth is represented by the variable *shallow*, which equals 1/(depth of profile midpoint). The *Yunnan* variables adjust for the fact that the Yunnan soil profile data do not specify whether the site was tilled. In equations (1)–(5) for the south, the six *Yunnan* variables are a *Yunnan* binary and the five basic soil chemistry parameters each multiplied by the *Yunnan* binary. For equations (6)–(10) in the south, the four *Yunnan* variables are the binary plus that binary times each of three logged agricultural variables (gross value of agriculture output, fertilizer, and the multiple-cropping index).

Regression type

2SLS—two-stage least squares; tobit—tobit analysis of a dependent variable bounded by zero.

In the tobit regressions for alkalinity and acidity, the coefficients shown here are the unnormalized coefficients times the share of cases with the dependent variable above zero. The part of R-squared is played by the correlation between the expected and observed values of alkalinity. Mean square error stands in for the standard error of estimate, and the part of the dependent-variable mean is played by the expected value of alkalinity for mean values of all independent variables.

Table C.2
Determinants of soil chemistry in cultivated soils of China, 1980s

North China

	Equation numbers and dependent variables							
	(6) ln (percentage organic matter)		(7) ln (total nitrogen)		(9) ln (total potassium)		(10) Alkalinity = max (pH − 8, 0)	
Independent variables	Coeff.	|t|	Coeff.	|t|	Coeff.	|t|	Coeff.	|t|
Endogenous								
ln (main crops/cultivated area)	1.010	(0.70)	0.891	(0.63)			0.501	(0.91)
ln (main crops/cultivated area), squared	−0.051	(0.55)	−0.056	(0.62)			−0.041	(1.14)
ln(all output/main crops)	0.709	(3.54)	0.489	(2.51)			−0.023	(0.23)
ln (all output/main crops), squared	−0.173	(2.61)	−0.156	(2.42)			−0.031	(0.20)
ln (multiple-cropping index)	−0.317	(0.43)	−0.349	(0.49)			−0.309	(1.30)
ln (multiple-cropping index), squared	0.357	(0.33)	−0.008	(0.01)			0.307	(0.85)
ln (fertilizer/cultivated area)	−0.118	(1.29)	0.009	(0.10)			0.0009	(0.02)
ln (fertilizer/cultivated area), squared	0.0005	(0.02)	0.027	(1.08)			−0.007	(0.60)
Exogenous								
Irrigated area/ cultivated area	0.022	(0.11)	0.219	(1.17)			0.041	(0.61)
Number of other exogenous variables	44		44				44	

Table C.2
(continued)

South China

Equation numbers and dependent variables

Independent variables	(6) ln (percentage organic matter)		(7) ln (total nitrogen)		(9) ln (total potassium)		(10) Acidity = max (6.0 − pH, 0)	
	Coeff.	$|t|$	Coeff.	$|t|$	Coeff.	$|t|$	Coeff.	$|t|$
Adjusted R-squared	0.578		0.470				0.582	
Dependent variable mean	0.249		−2.526				0.0001	
Standard error of estimate	0.385		0.375				0.021	
Regression type	2SLS		2SLS				2-stage tobit	
Endogenous								
ln (main crops/cultivated area)	1.888	(0.63)	−0.169	(0.06)	−0.219	(0.06)	7.227	(2.60)
ln (main crops/cultivated area), squared	−0.144	(0.80)	−0.036	(0.21)	0.016	(0.07)	−0.439	(2.65)
ln (all output/main crops)	−0.115	(0.35)	−0.092	(0.30)	0.693	(1.72)	−0.181	(0.58)
ln (all output/main crops), squared	0.654	(1.63)	0.387	(1.02)	−0.953	(1.93)	−0.149	(0.39)
ln (multiple-cropping index)	0.356	(0.42)	0.791	(0.99)	1.448	(1.40)	−0.772	(1.05)
ln (multiple-cropping index), squared	−0.397	(0.66)	−0.578	(1.01)	−1.332	(1.79)	0.686	(1.31)
ln (fertilizer/cultivated area)	−0.049	(0.69)	−0.064	(0.94)	0.170	(1.93)	−0.033	(0.56)
ln (fertilizer/cultivated area), squared	−0.004	(0.12)	−0.011	(0.35)	−0.034	(0.88)	0.022	(0.75)

Table C.2
(continued)

Independent variables	Equation numbers and dependent variables							
	(6) ln (percentage organic matter)		(7) ln (total nitrogen)		(9) ln (total potassium)		(10) Acidity = max (6.0 – pH, 0)	
	Coeff.	|t|	Coeff.	|t|	Coeff.	|t|	Coeff.	|t|
Exogenous								
Irrigated area/ cultivated area	0.007	(3.18)	0.008	(3.93)	0.002	(0.96)	0.002	(0.93)
Number of other exogenous variables	43		43		43		43	
Adjusted R-squared	0.382		0.370		0.364		0.287	
Dependent variable mean	0.894		−2.014		0.318		0.224	
Standard error of estimate	0.459		0.434		0.566		0.179	
Regression type	2SLS		2SLS		2SLS		2-stage tobit	

Notes: See note to table C.1 for samples, dependent variables, independent variables, and regression types.

Two other equations, eq. (8) for the log of total phosporus and Eq. (9) for the log of total potassium, proved insignificant as whole equations for north China. Phosporus and potassium content are therefore treated as exogenous variables for the north.

Eq. (8) for the log of total phosporus, proved insignificant as a whole equation for south China. Phosporus content is therefore treated as an exogenous variable for the south.

Appendix D: Full Regression Equation Estimates for the Determinants of Soil Chemistry in Indonesia, 1923–1990

Table D.1
Full regression equations for determinants of soil chemistry characteristics in Java, 1923–1990

Independent variables	Equation (1): (ln of % of) organic matter		Equation (2): (ln of % of) total nitrogen		Equation (3): (ln of mg/100g of) total P_2O_5		Equation (4): (ln of mg/100g of) total K_2O		Equation (5): max $(6 - pH, 0)$ = acidity	
	Coeff.	$|t|$	Coeff.	$|t|$	Coeff.	$|t|$	Coeff.	$|t|$	Coeff.	$|t|$
Fixed effects of land use type (relative to paddy)										
Tegalan (nonpaddy field crops)	0.1482	(0.56)	−0.1230	(0.20)	−0.0628	(0.38)	−0.1735	(1.16)	−0.1472	(0.89)
Tree crops	0.2379	(1.24)	−0.0048	(0.01)	−0.2698	(1.52)	−0.1696	(1.06)	0.2545	(1.61)
Fallow (grasses, bush, ladang)	0.1990	(0.93)	0.2113	(0.40)	−0.4740	(2.46)*	−0.1632	(0.94)	0.1641	(0.97)
Primary forest	0.3476	(1.48)	1.0033	(1.60)	−0.5249	(2.54)*	0.2350	(1.26)	−0.2519	(1.14)
No land use given	0.2727	(1.57)	0.5642	(1.15)	−0.1532	(1.10)	−0.1108	(0.89)	0.0811	(0.62)
Spline function time slopes by land use type (for cumulative trend results, see table 7.4)										
Post–World War II, any use	−0.1107	(1.07)	0.8818	(1.86)[a]	−0.0950	(0.74)	0.1389	(1.21)	0.3392	(2.71)**
Data 1974–1986, any use	0.0984	(1.11)	−0.6791	(4.09)**	−0.5225	(3.65)**	−0.4660	(3.65)**	0.2535	(2.16)*
Paddy, number of years before 1940	−0.1454	(1.76)[b]	0.1027	(1.37)						
Paddy, number of years since 1947	−0.0181	(0.57)	−0.1068	(2.58)*	0.0214	(0.60)	−0.0599	(1.86)[a]	−0.0425	(1.18)
Paddy, number of years since 1955	0.0182	(0.25)	0.1533	(1.98)[a]	−0.0443	(0.45)	0.0599	(0.67)	0.0752	(0.76)

Table D.1
(continued)

	Equation numbers and dependent variables									
	Equation (1): (ln of % of) organic matter		Equation (2): (ln of % of) total nitrogen		Equation (3): (ln of mg/100g of) total P_2O_5		Equation (4): (ln of mg/100g of) total K_2O		Equation (5): max $(6 - pH, 0) =$ acidity	
Independent variables	Coeff.	$|t|$	Coeff.	$|t|$	Coeff.	$|t|$	Coeff.	$|t|$	Coeff.	$|t|$
Paddy, number of years since 1960	-0.0155	(0.26)	-0.0645	(1.12)	0.0163	(0.20)	0.0094	(0.13)	-0.1104	(1.34)
Paddy, number of years since 1970	0.0203	(0.95)	0.0165	(0.86)	0.0299	(0.99)	-0.0003	(0.01)	0.1259	(3.81)**
Tegalan, number of years since 1947	-0.0390	(1.43)	-0.1098	(1.67)[b]	0.0542	(2.39)*	-0.0158	(0.77)	-0.0559	(2.27)*
Tegalan, number of years since 1955	0.0489	(1.12)	0.1968	(2.56)*	-0.1073	(2.23)*	0.0432	(1.00)	0.1027	(1.87)[a]
Tegalan, number of years since 1960	-0.0191	(0.61)	-0.1118	(3.60)**	0.0787	(1.82)[a]	-0.0246	(0.63)	-0.0863	(1.79)[b]
Tegalan, number of years since 1970	0.0060	(0.46)	0.0213	(1.75)[b]	-0.0341	(1.84)[a]	0.0014	(0.08)	0.0619	(3.09)**
Tree crops, number of years before 1940	-0.0782	(1.21)	0.1412	(2.40)*						
Tree crops, number of years since 1947	-0.0135	(0.73)	-0.0875	(1.65)[b]	0.0723	(3.02)**	0.0074	(0.34)	-0.1289	(5.25)**
Tree crops, number of years since 1955	-0.0047	(0.12)	0.1100	(1.70)[b]	-0.1505	(2.86)**	-0.0652	(1.37)	0.2528	(4.42)**

Table D.1
(continued)

Independent variables	Equation (1): (ln of % of) organic matter		Equation (2): (ln of % of) total nitrogen		Equation (3): (ln of mg/100g of) total P_2O_5		Equation (4): (ln of mg/100g of) total K_2O		Equation (5): max $(6 - pH, 0)$ = acidity	
	Coeff.	|t|	Coeff.	|t|	Coeff.	|t|	Coeff.	|t|	Coeff.	|t|
Tree crops, number of years since 1960	−0.0060	(0.18)	−0.0409	(1.31)	0.0972	(2.07)*	0.0716	(1.68)[b]	−0.1575	(3.16)**
Tree crops, number of years since 1970	0.0369	(1.72)[b]	0.0245	(1.29)	0.0142	(0.48)	−0.0158	(0.59)	0.0643	(2.22)*
Fallow, number of years before 1940	−0.1448	(1.03)	0.0695	(0.55)						
Fallow, number of years since 1947	−0.0234	(1.00)	−0.1082	(1.84)[a]	0.0702	(2.42)*	−0.0178	(0.68)	−0.0483	(1.77)[b]
Fallow, number of years since 1955	0.0291	(0.57)	0.0985	(1.33)	−0.0923	(1.35)	−0.0048	(0.08)	0.0070	(0.11)
Fallow, number of years since 1960	−0.0315	(0.73)	−0.00001	(0.00)	0.0303	(0.50)	0.0681	(1.23)	0.0274	(0.46)
Fallow, number of years since 1970	0.0354	(1.54)	0.0036	(0.17)	−0.0020	(0.06)	−0.0239	(0.76)	0.0193	(0.55)
Primary forest, since 1947	−0.0859	(2.52)*	−0.2863	(3.79)**	0.0406	(0.95)	−0.0674	(1.75)[b]	0.0460	(1.05)
Primary forest, since 1955	0.3734	(3.54)**	0.5776	(4.85)**	0.1071	(0.83)	0.1028	(0.88)	−0.0943	(0.61)
Primary forest, since 1960	−0.1799	(1.37)	−0.3157	(4.31)**	−0.1404	(1.28)	−0.0004	(0.00)	−0.0619	(0.32)
Primary forest, since 1970	−0.2258	(1.89)[a]	0.1422	(0.87)						

Table D.1
(continued)

Independent variables	Equation (1): (ln of % of) organic matter		Equation (2): (ln of % of) total nitrogen		Equation (3): (ln of mg/100g of) total P_2O_5		Equation (4): (ln of mg/100g of) total K_2O		Equation (5): max $(6 - pH, 0)$ = acidity	
	Coeff.	|t|	Coeff.	|t|	Coeff.	|t|	Coeff.	|t|	Coeff.	|t|
No land use given, number of years pre-1940	0.0121	(1.14)	−0.0944	(3.43)**	−0.0022	(0.25)	−0.0170	(2.08)*	0.0315	(3.15)**
No land use given, number of years since 1947	−0.0387	(1.55)	−0.1719	(2.87)**	0.0873	(2.69)**	−0.0380	(1.37)	−0.0175	(0.56)
No land use given, number of years since 1955	0.0409	(0.64)	0.1743	(1.71)[b]	−0.3233	(3.47)**	0.0424	(0.53)	−0.0409	(0.40)
No land use given, number of years since 1960	−0.0801	(1.22)	−0.0078	(0.11)	0.2895	(3.31)**	0.0119	(0.15)	0.0962	(0.77)
No land use given, number of years since 1970	0.1234	(2.18)*	−0.0120	(0.14)						
Effects of texture and topsoil depth										
Sand %	0.0361	(4.66)**	0.0374	(4.74)**	0.0502	(4.87)**	0.0398	(4.27)**	0.0040	(0.32)
Sand % squared	−0.0005	(7.49)**	−0.0005	(6.61)**	−0.0005	(5.33)**	−0.0004	(4.98)**	−0.0001	(1.09)
Clay %	−0.0042	(0.61)	−0.0007	(0.11)	0.0176	(1.82)[b]	0.0263	(3.01)**	−0.0013	(0.12)
Clay % squared	0.0001	(1.04)	0.00004	(0.66)	−0.0002	(2.17)*	−0.0002	(3.19)**	0.0001	(0.92)
Sand % times clay %	−0.0006	(4.54)**	−0.0007	(5.49)**	−0.0008	(4.56)**	−0.0007	(4.22)**	−0.0002	(0.86)
No texture given	0.0990	(0.43)	0.4139	(1.83)[a]	0.7578	(2.22)*	0.3785	(1.22)	0.1452	(0.39)
Shallowness of topsoil	0.9793	(3.84)**	0.6713	(2.55)	−0.9134	(2.66)**	0.6202	(1.99)[a]	−0.0154	(0.04)

Table D.1
(continued)

Independent variables	Equation (1): (ln of % of organic matter)		Equation (2): (ln of % of total nitrogen)		Equation (3): (ln of mg/100g of total P_2O_5)		Equation (4): (ln of mg/100g of total K_2O)		Equation (5): max $(6 - pH, 0)$ = acidity	
	Coeff.	|t|	Coeff.	|t|	Coeff.	|t|	Coeff.	|t|	Coeff.	|t|
Effects of terrain (relative to flat land)										
Sloped land	0.0183	(0.40)	0.0728	(1.72)[b]	−0.0943	(1.52)	−0.1334	(2.37)*	−0.0029	(0.05)
Undulating	0.0928	(1.22)	0.1757	(2.54)*	−0.2342	(2.30)*	−0.1573	(1.70)[b]	0.2008	(2.09)*
Hilly or steep terrain	0.0263	(0.45)	0.0840	(1.48)	−0.1490	(1.99)[a]	−0.1424	(2.10)*	0.1047	(1.43)
Other terrain descriptions	0.0499	(0.75)	0.1634	(2.59)*	0.1036	(1.16)	−0.0578	(0.71)	0.2449	(2.83)**
No terrain description given	−0.0308	(0.62)	0.0002	(0.00)	0.0884	(1.52)	−0.0359	(0.68)	−0.2086	(3.48)**
Effects of parent material (geology classes, relative to alluvial)										
Pleistocene	0.0515	(0.91)	0.0679	(1.14)	0.1891	(2.82)**	−0.1429	(2.35)*	0.1835	(2.65)**
Plio-pleistocene, Pliocene	0.3625	(3.07)**	0.0339	(0.23)	0.4411	(3.57)**	−0.0667	(0.60)	0.3423	(2.65)**
Mio-pliocene, Upper Miocene	−0.0090	(0.12)	0.0401	(0.49)	−0.2623	(2.92)**	−0.0503	(0.62)	0.2773	(3.10)**
Paleogene thru Jurassic	0.1274	(0.68)	−0.2923	(0.58)	0.0316	(0.14)	0.0161	(0.08)	0.0563	(0.26)
Recent (alluvial facies)	−0.1885	(1.60)	−0.3720	(2.52)*	0.2553	(1.69)[b]	0.4250	(3.10)**	−0.1653	(1.02)
Other recent	0.0857	(1.51)	0.1085	(1.87)[a]	0.1222	(1.65)[b]	−0.2720	(4.05)**	0.3406	(4.63)**
Younger quaternary	−0.0308	(0.48)	−0.0363	(0.54)	0.1034	(1.25)	0.0510	(0.67)	−0.0809	(0.84)
Old quaternary, intermediate basic	0.0587	(0.57)	0.1980	(1.84)[a]	−0.0232	(0.17)	−0.2332	(1.86)[a]	−0.1425	(0.89)
Mediterranean suite	0.1864	(1.14)	0.1445	(0.60)	0.0090	(0.04)	0.0334	(0.17)	0.3867	(1.89)[a]

Table D.1
(continued)

| | Equation numbers and dependent variables | | | | | | | | | |
| | Equation (1): (ln of % of organic matter) | | Equation (2): (ln of % of total nitrogen) | | Equation (3): (ln of mg/100g of total P_2O_5) | | Equation (4): (ln of mg/100g of total K_2O) | | Equation (5): max $(6 - pH, 0)$ = acidity | |
Independent variables	Coeff.	$|t|$	Coeff.	$|t|$	Coeff.	$|t|$	Coeff.	$|t|$	Coeff.	$|t|$
Effects of precipitation (relative to 5–6 months wet, 2–3 dry)										
> 9 months wet, < 2 dry	0.2534	(3.69)**	0.2989	(3.81)**	0.4793	(5.20)**	−0.2918	(3.49)**	0.7953	(9.06)**
7–9 months wet, < 2 dry	0.2644	(4.30)**	0.1921	(3.01)**	0.3054	(3.78)**	−0.2456	(3.36)**	0.6548	(8.38)**
7–9 months wet, 2–4 dry	−0.0849	(1.16)	0.0059	(0.07)	0.2029	(2.31)*	−0.1186	(1.47)	0.2670	(3.06)**
5–6 months wet, 3–5 dry	−0.0896	(1.47)	−0.1778	(2.80)**	−0.0080	(0.11)	−0.1283	(1.96)[a]	−0.0270	(0.32)
3–4 months wet, 2–3 dry	−0.0348	(0.54)	0.0334	(0.46)	−0.1065	(1.35)	−0.1349	(1.87)[a]	0.4016	(5.10)**
3–4 months wet, 3–5 dry	0.0017	(0.03)	−0.0113	(0.22)	−0.0859	(1.37)	0.0613	(1.08)	0.1359	(1.99)[a]
< 3 months wet	0.0901	(1.16)	0.1747	(2.01)*	0.3189	(3.21)**	0.3107	(3.50)**	0.2224	(2.12)*
Effects of soil class (relative to soil class 10)										
Hydraquents	0.1240	(0.91)	0.1819	(1.46)	0.0080	(0.04)	0.3228	(1.74)[b]	0.0198	(0.10)
Classes 4, 5, 18 (Lindert 1997, table A4)	−0.8165	(7.56)**	−0.3165	(1.49)	−0.5194	(4.57)**	−0.4397	(4.24)**	0.0910	(0.72)
Tropopsamments	0.0098	(0.05)	−0.2497	(0.90)	−0.0898	(0.39)	−0.2968	(1.47)	0.3274	(1.48)
Dystrandepts	0.1131	(1.66)[b]	0.1783	(2.54)*	−0.0914	(1.03)	0.0589	(0.73)	0.0327	(0.37)
Eutrandepts, vitrandepts	0.7493	(3.97)**	1.2795	(2.59)*	0.3596	(1.20)	−0.2543	(0.93)	0.1096	(0.45)
Dystropepts	0.0557	(0.85)	0.0698	(0.98)	−0.0429	(0.53)	−0.1601	(2.18)*	0.2949	(3.59)**
Eutropepts	−0.1324	(2.48)*	−0.0432	(0.75)	−0.1987	(2.79)**	−0.2071	(3.19)**	0.0278	(0.36)
Ustropepts	−0.0455	(0.53)	−0.0995	(1.10)	0.2534	(2.34)*	−0.0792	(0.80)	0.1296	(1.07)
Classes 19–21 (Lindert 1997, T. A4)	0.1053	(0.50)	−0.0441	(0.20)	−0.6676	(2.60)*	0.0000	(0.00)	0.1458	(0.55)

Table D.1
(continued)

Independent variables	Equation (1): (ln of % of) organic matter		Equation (2): (ln of % of) total nitrogen		Equation (3): (ln of mg/100g of) total P_2O_5		Equation (4): (ln of mg/100g of) total K_2O		Equation (5): max $(6 - pH, 0)$ = acidity	
	Coeff.	|t|	Coeff.	|t|	Coeff.	|t|	Coeff.	|t|	Coeff.	|t|
Calciorthids	0.0730	(0.40)	0.1830	(1.18)	0.021919	(0.08)	−0.8522	(3.66)**	0.8644	(3.59)**
Tropudolfs	0.1333	(2.08)*	0.1441	(2.13)*	−0.1520	(1.88)[a]	−0.1977	(2.69)**	0.2406	(2.90)**
Paleustalfs	0.4096	(2.64)**	0.6011	(4.45)**	1.2575	(3.41)**	0.9907	(3.26)**	0.3631	(1.81)[b]
Haplustalfs	−0.2841	(3.04)**	−0.0223	(0.20)	−0.6720	(5.78)**	−0.0845	(0.81)	−0.1418	(1.08)
Paleudults	−0.0590	(0.58)	0.0132	(0.13)	0.0354	(0.25)	−0.4909	(3.80)**	0.2054	(1.58)
Tropudults	0.1593	(1.81)[b]	0.2493	(2.89)**	0.0650	(0.60)	−0.2543	(2.59)*	0.3863	(3.76)**
Paleustults	0.4034	(2.16)*	0.1360	(0.68)	−0.1281	(0.74)	0.0836	(0.54)	−0.6105	(2.37)*
Soil class unknown	0.0987	(0.63)	0.2082	(1.00)	−0.0939	(0.59)	−0.1646	(1.10)	−0.3182	(1.70)[b]
Separate effects from districts (kabupaten) having peculiar macronutrient patterns at all times										
Jakarta municipality	0.0624	(0.58)	0.8078	(3.07)**	0.1495	(1.01)	−0.6238	(4.62)**	0.5133	(3.85)**
Sukabumi	0.0940	(0.49)	−0.6692	(2.10)*	0.1437	(0.75)	−0.2359	(1.36)	−0.4590	(2.55)*
Cianjur	0.1948	(2.04)*	0.0220	(0.22)	−0.3472	(2.60)**	−0.3613	(3.02)**	0.0675	(0.56)
Garut	0.6290	(5.87)**	0.1515	(0.92)						
Tangerang	−0.1663	(1.14)	0.1228	(1.39)	−0.2070	(0.99)	−0.9140	(4.82)**	0.6589	(4.06)**
Purbalingga	1.0036	(4.83)**	0.5215	(5.42)**						
Banjarngara	0.5372	(2.58)*	−0.2433	(1.62)						
Sukohardjo	−0.0246	(0.20)	1.2553	(6.27)**	−0.5917	(3.45)**	−0.8252	(5.39)**	0.4004	(2.17)*
Pati	−0.0735	(0.93)	0.6444	(2.78)**	1.2364	(11.94)**	0.5215	(5.58)**	0.2405	(2.20)*
Jepara	0.1172	(0.58)	−0.2604	(2.00)[a]	1.0045	(3.64)**	0.4699	(1.93)[a]	0.4025	(1.54)
Demak	0.2627	(3.09)**	−0.0870	(1.16)	0.2970	(2.51)*	0.6718	(6.25)**	−3.6706	(0.00)

Table D.1
(continued)

Independent variables	Equation (1): (ln of % of) organic matter		Equation (2): (ln of % of) total nitrogen		Equation (3): (ln of mg/100g of) total P₂O₅		Equation (4): (ln of mg/100g of) total K₂O		Equation (5): max (6 − pH, 0) = acidity	
	Coeff.	|t|	Coeff.	|t|	Coeff.	|t|	Coeff.	|t|	Coeff.	|t|
Banyuwangi	0.5557	(2.75)**	0.4626	(1.71)[b]	0.1576	(0.55)	0.9929	(3.79)**	−0.2610	(0.64)
Pasaruan	−0.3864	(2.81)**	0.2652	(3.44)**	0.0970	(0.59)	1.0375	(6.93)**	−0.1510	(0.74)
Magetan	−0.3739	(2.21)*	0.7242	(3.48)**	−0.7374	(3.05)**	−0.7779	(3.53)**	−0.2566	(0.87)
Surabaya municipality	−0.2915	(2.61)*	−0.3390	(2.52)*	0.2865	(1.81)[b]	0.5187	(3.60)**	−1.0007	(2.77)**
Bali	0.7119	(3.21)**	0.2743	(1.41)	1.0183	(5.74)**	−0.6716	(1.96)[a]		
Constant	0.9249	(3.34)**	−2.3512	(4.27)**	3.2865	(10.02)**	3.1721	(10.73)**	−0.5370	(1.41)
Adjusted R-squared	0.4036		0.4173		0.2808		0.3760		0.4267	
Dependent variable mean	0.9602		−1.9991		3.8130		3.5709		0.1583	
Standard error of estimate	0.5428		0.4742		0.7799		0.7092		0.6716	
Regression type	OLS		OLS		OLS		OLS		tobit	
Number of observations	1,777		1,302		2,115		2,126		2,191	

Notes: Statistical significance (two-tailed): **1-percent level, *5-percent level, [a]6-percent level, [b]7-percent level. For the tobit equation, the dependent variable is *Acid* = max (6 − pH, 0). In the acidity equation here, the share of observations with acidity above zero is .4135. Its R-squared is the squared correlation between observed and expected values. In the acidity equation here, the share of observations with acidity above zero is .4135. Multiplying by this decimal fraction converts each regression coefficient into a predicted overall effect at mean values of the independent variables. On the tobit regression approach to dealing with bounded dependent variables (e.g., bounded by zero, as here), see Maddala 1983. As partially implied by the headings over independent variables, the base case to which others are compared is an alluvial tropaquept in flat lowland paddy with five to six months wet and two to three months dry around the year 1940. Such a soil profile might have been taken near Semerang in Central Java. A "wet" month averages more than 200 mm of precipitation. A "dry" month averages less than 100 mm.

Table D.2
Full regression equations for determinants of soil chemistry characteristics on the Outer Islands, 1923–1990

Independent variables	Equation (1): (ln of % of) organic matter Coeff.	$\|t\|$	Equation (2): (ln of % of) total nitrogen Coeff.	$\|t\|$	Equation (3): (ln of mg/100g of) total P_2O_5 Coeff.	$\|t\|$	Equation (4): (ln of mg/100g of) total K_2O Coeff.	$\|t\|$	Equation (5): max $(6 - pH, 0)$ = acidity Coeff.	$\|t\|$
Fixed effects of land use type (relative to paddy)										
Tegalan (nonpaddy field crops)	0.5564	(1.57)	0.6308	(1.92)[a]	−2.4209	(4.31)**	−1.4821	(2.62)*	0.2393	(0.47)
Tree crops	0.9691	(2.65)**	−0.2558	(0.75)	−0.5326	(1.07)	−0.7583	(1.52)	0.1930	(0.43)
Fallow (grasses, bush, *ladang*)	0.1557	(0.46)	−0.2092	(0.53)	−0.7981	(1.65)[b]	−0.8232	(1.69)[b]	0.1585	(0.35)
Primary forest	0.3128	(1.29)	0.4167	(1.79)[b]	−0.7755	(1.55)	−0.5412	(1.08)	0.2616	(0.56)
No land use given	0.8250	(3.02)**	−0.2384	(0.89)	−0.6569	(1.36)	−0.8417	(1.73)[b]	0.3206	(0.72)
Spline function time trends by land use type (for cumulative trend results, see table 7.4)										
Post–World War II, any use	0.3990	(1.50)	−0.6600	(2.59)*	−0.3180	(2.37)*	−0.3749	(2.77)**	−0.2183	(1.51)
Data 1974–1986, any use	−0.1443	(1.60)			−0.6328	(6.91)**	−0.3801	(3.93)**	−0.0163	(0.23)
Paddy, number of years since 1947					−0.2098	(1.66)[b]	0.1039	(0.83)	0.3864	(3.31)**
Paddy, number of years since 1955					0.4892	(1.33)	−0.4174	(1.14)	−1.1786	(3.44)**
Paddy, number of years since 1960	0.0020	(0.23)	0.0125	(1.56)	−0.2952	(0.99)	0.3376	(1.13)	0.9092	(3.24)**
Paddy, number of years since 1970					0.0515	(0.81)	−0.0832	(1.31)	−0.1321	(2.18)*

Equation numbers and dependent variables

	Equation numbers and dependent variables									
	Equation (1): (ln of % of) organic matter		Equation (2): (ln of % of) total nitrogen		Equation (3): (ln of mg/100g of) total P_2O_5		Equation (4): (ln of mg/100g of) total K_2O		Equation (5): max $(6 - pH, 0)$ = acidity	
Independent variables	Coeff.	|t|	Coeff.	|t|	Coeff.	|t|	Coeff.	|t|	Coeff.	|t|
Tegalan, number of years since 1947	-0.0133	(1.37)	-0.0130	(1.45)	0.1652	(4.76)**	0.0926	(2.65)**	-0.0493	(0.99)
Tegalan, number of years since 1955									0.0665	(1.12)
Tegalan, number of years since 1960					-0.1545	(1.79)[b]	-0.0643	(0.75)		
Tegalan, number of years since 1970					-0.0323	(0.33)	-0.0906	(0.94)		
Tree crops, number of years before 1940					-0.1800	(1.77)[b]	0.1854	(1.80)[b]		
Tree crops, number of years since 1947	-0.0946	(2.14)*	0.0425	(1.02)	-0.0082	(0.34)	0.0421	(1.71)[b]	0.0905	(3.83)**
Tree crops, number of years since 1955	0.1335	(1.92)[a]	0.0042	(0.06)	0.0066	(0.12)	-0.0135	(0.24)	-0.2333	(4.27)**
Tree crops, number of years since 1960	-0.0762	(1.65)[b]	-0.0726	(1.71)[b]	0.0205	(0.40)	-0.0690	(1.35)	0.1643	(3.27)**
Tree crops, number of years since 1970	0.0450	(2.42)*	0.0294	(1.73)[b]	-0.0275	(1.31)	0.0098	(0.46)	0.0015	(0.07)

Table D.2
(continued)

Independent variables	Equation (1): (ln of % of) organic matter		Equation (2): (ln of % of) total nitrogen		Equation (3): (ln of mg/100g of) total P_2O_5		Equation (4): (ln of mg/100g of) total K_2O		Equation (5): max $(6 - pH, 0)$ = acidity	
	Coeff.	$|t|$	Coeff.	$|t|$	Coeff.	$|t|$	Coeff.	$|t|$	Coeff.	$|t|$
Fallow, number of years since 1947	0.0106	(0.27)	0.0374	(0.81)	0.0357	(1.70)	0.0620	(2.92)**	0.0552	(2.49)*
Fallow, number of years since 1955	−0.0259	(0.44)	0.0081	(0.13)	−0.1254	(2.67)**	−0.1072	(2.26)*	−0.0708	(1.53)
Fallow, number of years since 1960	−0.0011	(0.03)	−0.0716	(1.90)[a]	0.1218	(2.79)**	0.0457	(1.04)	−0.0124	(0.29)
Fallow, number of years since 1970	0.0147	(0.76)	0.0268	(1.51)	−0.0302	(1.39)	−0.0379	(1.72)[b]	0.0576	(2.76)**
Primary forest, since 1947	−0.0057	(0.89)	−0.0061	(0.99)	0.0012	(0.24)	−0.0111	(2.15)*	0.0194	(4.14)**
No land use given, number of years pre-1940	−0.0264	(0.69)	−0.0065	(0.35)	0.0140	(1.03)	−0.0148	(1.08)	−0.0234	(1.55)
No land use given, number of years since 1947	−0.0746	(1.95)[a]	0.0581	(1.54)	−0.0274	(0.83)	0.0352	(1.06)	0.0815	(2.46)*
No land use given, number of years since 1955	0.0572	(1.27)	−0.0735	(1.67)[b]	0.1397	(1.12)	0.0994	(0.79)	−0.2434	(2.04)*
No land use given, number of years since 1960					−0.1349	(0.85)	−0.2674	(1.67)[b]	0.1833	(1.23)
No land use given, number of years since 1970					0.0210	(0.23)	0.1417	(1.57)	0.0005	(0.01)

Table D.2
(continued)

	Equation numbers and dependent variables									
	Equation (1): (ln of % of) organic matter		Equation (2): (ln of % of) total nitrogen		Equation (3): (ln of mg/100g of) total P_2O_5		Equation (4): (ln of mg/100g of) total K_2O		Equation (5): max $(6 - pH, 0)$ = acidity	
Independent variables	Coeff.	|t|	Coeff.	|t|	Coeff.	|t|	Coeff.	|t|	Coeff.	|t|
Effects of texture and topsoil depth										
Sand %	−0.0167	(2.85)**	−0.0112	(2.03)*	−0.0110	(1.65)[b]	−0.0008	(0.12)	−0.0034	(0.54)
Sand % squared	0.0001	(1.41)	0.0000	(0.42)	0.0000	(0.30)	−0.0001	(1.25)	0.0000	(0.25)
Clay %	−0.0129	(1.93)[a]	−0.0021	(0.33)	0.0035	(0.46)	0.0200	(2.66)**	−0.0085	(1.16)
Clay % squared	0.0001	(1.99)[a]	0.0000	(0.55)	−0.0001	(1.10)	−0.0003	(4.52)**	0.0001	(2.14)*
Sand % times clay %	0.0002	(1.83)[a]	0.0001	(0.62)	−0.0003	(2.43)*	−0.0002	(1.91)[a]	0.0002	(1.73)[b]
No texture given	−0.7816	(3.69)**	−0.4266	(2.11)*	−0.7645	(3.14)**	−0.5742	(2.41)*	0.1632	(0.68)
Shallowness of topsoil	2.0416	(7.19)**	1.4005	(5.26)**	1.1548	(3.93)**	1.2928	(4.44)**	0.4784	(1.72)[b]
Effects of terrain (relative to flat land)										
Sloped land	−0.0604	(1.06)	0.0259	(0.49)	−0.1431	(2.38)*	−0.0406	(0.67)	−0.0050	(0.09)
Undulating	0.0333	(0.47)	−0.0016	(0.02)	−0.2826	(3.49)**	0.0087	(0.11)	0.0025	(0.03)
Hilly or steep terrain	0.0031	(0.05)	0.0548	(1.06)	−0.1815	(3.08)**	−0.1022	(1.73)[b]	−0.0697	(1.23)
Other terrain descriptions	0.1726	(1.57)	0.1424	(1.39)	0.1837	(1.43)	0.1414	(1.10)	0.0811	(0.67)
No terrain description given	−0.0686	(1.02)	−0.0567	(0.91)	−0.0612	(0.92)	−0.0785	(1.17)	−0.0323	(0.51)
Effects of parent material (geology classes, relative to alluvial)										
Pleistocene	−0.0114	(0.03)	−0.0209	(0.06)	0.3762	(0.81)	−0.9508	(2.02)*	0.1122	(0.26)
Plio-pleistocene, pliocene	0.1058	(1.01)	0.1413	(1.44)	−0.0029	(0.03)	−0.5464	(4.75)**	0.2083	(1.94)[a]
Mio-pliocene, Upper Miocene	−0.2469	(1.96)[a]	0.0232	(0.20)	−0.0487	(0.38)	−0.2349	(1.82)[a]	−0.1158	(0.95)

Table D.2
(continued)

Independent variables	Equation (1): (ln of % of) organic matter		Equation (2): (ln of % of) total nitrogen		Equation (3): (ln of mg/100g of) total P_2O_5		Equation (4): (ln of mg/100g of) total K_2O		Equation (5): max $(6 - pH, 0)$ = acidity	
	Coeff.	$\|t\|$	Coeff.	$\|t\|$	Coeff.	$\|t\|$	Coeff.	$\|t\|$	Coeff.	$\|t\|$
Middle Miocene through Neogene	-0.0432	(0.62)	-0.0712	(1.12)	-0.1041	(1.64)	-0.2802	(4.28)**	-0.0249	(0.41)
Paleogene through Jurassic	0.0312	(0.45)	-0.0111	(0.17)	-0.1446	(1.99)[a]	-0.2129	(2.91)**	0.1265	(1.91)[a]
Sedimentary classes 14–19	0.0112	(0.12)	-0.2114	(2.48)*	-0.1255	(1.25)	-0.4868	(4.76)**	-0.0588	(0.61)
Recent nonalluvial volcanic	0.3527	(4.63)**	0.2330	(3.27)**	0.1710	(2.34)*	-0.2251	(3.06)**	-0.0993	(1.41)
Old quaternary, intermediate basic	-0.1342	(1.37)	-0.3534	(3.75)**	0.0123	(0.13)	-0.5561	(5.67)**	-0.1040	(1.07)
Silic effusives, old andesite	0.2763	(2.32)*	0.1040	(0.84)	0.0758	(0.56)	0.3343	(2.48)*	0.1577	(1.24)
Mediterranean suite	-0.4954	(2.72)**	-0.1765	(1.02)	0.6836	(4.03)**	0.9009	(5.26)**	-0.8369	(4.74)**
Plutonic rocks	0.2098	(1.54)	-0.0714	(0.55)	-0.0709	(0.48)	-0.3131	(2.09)*	0.0831	(0.57)
Unknown geology class	-0.1651	(1.55)	-0.1838	(1.86)[a]	0.1537	(1.28)	0.1805	(1.72)[b]	-0.1586	(1.43)
Effects of precipitation (relative to 10–12 months wet, 0–2 dry)										
7–9 months wet, < 2 months dry	-0.0756	(1.20)	-0.2145	(3.58)**	-0.2848	(4.48)**	-0.3603	(5.61)**	0.3801	(6.49)**
5–6 months wet, < 2 months dry	-0.0387	(0.33)	0.0034	(0.03)	-0.0109	(0.11)	-0.2944	(3.06)**	0.0019	(0.02)

Table D.2
(continued)

Independent variables	Equation (1): (ln of % of) organic matter		Equation (2): (ln of % of) total nitrogen		Equation (3): (ln of mg/100g of) total P_2O_5		Equation (4): (ln of mg/100g of) total K_2O		Equation (5): max (6 − pH, 0) = acidity	
	Coeff.	|t|	Coeff.	|t|	Coeff.	|t|	Coeff.	|t|	Coeff.	|t|
5–6 months wet, 2–3 months dry	−0.1236	(1.95)[a]	−0.0726	(1.23)	−0.0828	(1.26)	−0.4322	(6.36)**	−0.0829	(1.45)
5–6 months wet, 3–5 months dry	−0.2450	(1.71)[b]	0.0271	(0.20)	0.2495	(2.08)*	−0.5834	(4.81)**	0.3488	(2.97)**
3–4 months wet, < 2 months dry	−0.0813	(0.90)	−0.2070	(2.47)*	−0.2779	(3.49)**	−0.4187	(5.12)**	0.4014	(5.30)**
3–4 months wet, 2–3 months dry	0.0180	(0.18)	0.0244	(0.27)	−0.3786	(3.52)**	−0.5590	(5.09)**	0.2093	(2.05)*
3–4 months wet, 3–5 months dry	−0.1563	(1.09)	−0.0330	(0.24)	0.1026	(0.74)	−0.2840	(2.03)*	−0.0424	(0.27)
3–4 months wet, > 5 months dry	−0.3277	(2.03)*	−0.0056	(0.04)	0.5964	(3.70)**	−0.0012	(0.01)	0.0883	(0.50)
<3 months wet, > 6 months dry	−0.1662	(2.03)*	−0.1426	(1.84)[a]	0.4230	(4.89)**	0.2232	(2.64)*	−0.0557	(0.65)
Effects of soil class (relative to soil class 10)										
Fluvaquents	−0.4313	(2.63)*	−0.0456	(0.27)	−0.0634	(0.35)	−0.0307	(0.17)	0.1075	(0.64)
Classes 2–6 (Lindert 1997, table A4)	−0.1112	(0.63)	−0.1677	(0.98)	−0.0988	(0.59)	0.4165	(2.44)*	−0.3532	(2.14)*

Table D.2
(continued)

Independent variables	Equation (1): (ln of % of) organic matter		Equation (2): (ln of % of) total nitrogen		Equation (3): (ln of mg/100g of) total P_2O_5		Equation (4): (ln of mg/100g of) total K_2O		Equation (5): max $(6 - pH, 0)$ = acidity	
	Coeff.	|t|	Coeff.	|t|	Coeff.	|t|	Coeff.	|t|	Coeff.	|t|
Dystrandepts	0.0841	(0.71)	0.1872	(1.71)[b]	0.3078	(2.40)*	0.2706	(2.09)*	-0.3153	(2.55)**
Eutrandepts, vitrandepts	0.5745	(2.17)*	0.3615	(1.41)	0.2594	(1.20)	0.4397	(2.03)*	0.1111	(0.45)
Dystropepts	-0.1457	(2.43)*	-0.0416	(0.74)	0.0782	(1.33)	0.0985	(1.65)[b]	-0.2254	(3.92)**
Eutropepts, ustropepts	0.2575	(1.88)[a]	0.3241	(2.54)*	0.0242	(0.18)	0.2150	(1.62)[b]	-0.5398	(3.49)**
Humitropepts, tropohemists	0.0241	(0.31)	0.0713	(0.99)	-0.2255	(2.92)**	-0.1849	(2.33)*	-0.0463	(0.64)
Classes 19–21 (1997, table A4)	0.1983	(1.37)	0.1084	(0.81)	0.1492	(1.16)	-0.1907	(1.46)	-0.4857	(3.39)**
Tropudolfs	0.0369	(0.30)	0.0003	(0.00)	-0.2207	(1.99)[a]	0.1512	(1.35)	-0.3958	(3.66)**
Rhodustalfs, Haplustalfs	-0.2096	(1.01)	0.0454	(0.23)	-0.1595	(0.65)	-0.7221	(2.93)**	0.4422	(1.70)[b]
Paleudults	-0.0094	(0.08)	-0.0604	(0.57)	-0.0721	(0.60)	-0.0375	(0.31)	-0.0324	(0.29)
Tropudults	-0.1689	(2.81)**	-0.0707	(1.26)	-0.2553	(4.20)**	-0.1055	(1.73)[b]	-0.0476	(0.86)
Soil classes 29–32 (Lindert 1997, table A4)	-0.1066	(0.46)	-0.2205	(1.00)	-0.3674	(2.08)*	-0.0880	(0.49)	-0.0452	(0.27)
Soil class unknown	-0.1608	(0.81)	0.1567	(0.86)	-0.2049	(0.92)	0.1354	(0.61)	-0.4609	(1.80)[b]
Separate effects from districts (kabupaten) having peculiar macronutrient patterns at all times										
Labuhan Batu, North Sumatra							1.0336	(2.62)*		

Table D.2
(continued)

Independent variables	Equation (1): (ln of % of) organic matter Coeff.	|t|	Equation (2): (ln of % of) total nitrogen Coeff.	|t|	Equation (3): (ln of mg/100g of) total P_2O_5 Coeff.	|t|	Equation (4): (ln of mg/100g of) total K_2O Coeff.	|t|	Equation (5): max $(6 - pH, 0)$ = acidity Coeff.	|t|
Ogan Komering Ilir, South Sumatra	0.3228	(2.42)*	0.1063	(0.87)	1.1176	(3.09)**	0.6301	(4.14)**		
Musi Banyuasin, South Sumatra	0.3084	(2.61)*	0.1311	(1.22)	-0.9582	(3.93)**	-0.5212	(4.32)**		
Belitung, South Sumatra							-1.1298	(3.40)**		
Kalimantan	0.2180	(3.46)**	0.1321	(2.24)*	0.0910	(1.49)	-0.3167	(4.86)**	0.1750	(3.11)**
South Sulawesi and Kolaka	-0.2534	(2.71)**	-0.0612	(0.69)	0.8918	(9.53)**	0.4092	(4.60)**	-0.2575	(3.02)**
Other Sulawesi	-0.1509	(1.59)	0.0634	(0.72)	0.1693	(1.92)[a]	0.7778	(8.15)**	-0.9488	(10.85)**
Maluku and Irian Jaya	0.2582	(1.65)[b]	0.4057	(2.79)**	0.7077	(5.88)**	0.6733	(5.53)**	-0.7909	(6.34)**
Nusa Tenggara	-0.5919	(4.37)**	-0.7568	(5.64)**	0.9276	(6.79)**	1.6692	(12.33)**	-1.6014	(9.98)**
Constant	1.2895	(3.86)**	-0.9124	(2.92)**	4.5574	(8.70)**	4.4828	(8.53)**	0.5559	(1.14)
Adjusted R-squared	0.2822		0.2839		0.4896		0.5253		0.4212	
Dependent var. mean	1.3927		-1.6590		3.2548		3.2797		0.6901	
Stand. error of estimate	0.6594		0.5942		0.7686		0.7750		0.7122	
Regression type	OLS		OLS		OLS		OLS		tobit	
Number of observations	1,703		1,613		2,176		2,240		2,312	

Table D.2
(continued)

Notes: Statistical significance (two-tailed): **—1-percent level, *—5-percent level, [a]—6-percent level, [b]—7-percent level. For the tobit equation, the dependent variable is *Acid* = max $(6 - pH, 0)$. Its R-squared is the squared correlation between observed and expected values. In the acidity equation here, the share of observations with acidity above zero is .7011. Multiplying by this decimal fraction converts each regression coefficient into a predicted overall effect at mean values of the independent variables. On the tobit regression approach to dealing with bounded dependent variables (e.g., bounded by zero, as here), see Maddala 1983. As partially implied by the headings over independent variables, the base case to which others are compared is an alluvial tropaquept in flat lowland paddy with 10–12 months wet and 0–2 months dry around the year 1940. Such a soil profile might have been taken near Kisaren in North Sumatra. A "wet" month averages more than 200 mm of precipitation. A "dry" month averages less than 100 mm.

Notes

Chapter 1

1. See also Orleans 1992; Wen 1993; and Brown 1995, chaps. 1, 4.

2. "Een groot gevaar, dat sluipend nadert," as cited in Donner 1987, p. 94.

3. World Resources Institute 1988, pp. 12 and 14. By contrast, the official estimates in table 2.3 imply that only 757,000 hectares—8 percent of agricultural farmland—were degraded in all of Java in 1987. As noted in chapter 2, none of these degraded-area estimates can establish a recent rate of erosion or a link to specific human activities.

4. Wen provides no footnote for these national aggregates. Two points relevant about his estimates for the issues discussed in chapter 2 are (1) that his figure of 150 million hectares must include considerable land that is not under human cultivation (see chapter 6), and (2) that the percentages are less than 100% of new nutrients applied, and a lower share of nutrients available in the topsoil.

Chapter 2

1. This is not the first study to criticize the contemporary literature's evidence. See Crosson 1985, 1995; Crosson and Stout 1983; Stocking 1987; Tiffen, Mortimore, and Gichuki 1991; and, on Java, Diemont, Smiet, and Nurdin 1991.

2. The "since 1945" is explicit in World Resources Institute 1992. They appear to base this division on a semiexplicit time threshold in the Explanatory Note to the original map, in which Oldeman, Hakkeling, and Sombroek (1991, Explanatory Note, p. 20) divide all history into three periods, the third being the "post second world war period." The note downplays the soil degradation of the two earlier periods, especially in a passage that cites "human-induced soil degradation in the past" (apparently before the third period) as one example of a number of aspects that "appeared to be of only minor importance" and so were "not represented on the map."

3. In its publications related to the GLASOD map, the International Soil Reference and Information Centre (ISRIC) describes the global assessment map as a

work in progress. True to this disclaimer, the work continues. A new mapping was prepared in 1994–1997 for China, departing in many details from the 1991 global map's impressions on China (van Lynden 1998). The related ISRIC project of accumulating detailed reference soil data from around the world can, with sufficient funding, become a powerful empirical tool in the twenty-first century, as chapter 3 suggests.

4. This danger of reading trends into the Ministry's degraded land area series does not significantly harm the results of Huang and Rozelle's (1995) use of the ministry's figures in their regression analysis of China's provincial yields per hectare. Huang and Rozelle's plausible estimates of the effects of land degradation rest mainly on the differences in degradation across provinces, not over time.

5. World Resources Institute 1988, p. 13. WRI cites an unpublished paper presented at the Earth as Transformed by Human Activities Conference, Worcester Massachusetts, October 1987, as its source for these data.

6. Specifically, Dregne (1982) estimated desertification shares of all lands on earth and converted them into estimates of losses in agricultural yields. He assumed that desertification was "slight" in 62 percent of the world's lands, "moderate" in 26 percent, "severe" in 12 percent, and "very severe" in 0.1 percent. These four classes were assumed to reduce yields by 0–10 percent, 10–50 percent, 50–90 percent, and more than 90 percent, respectively. For China, the areas he has mapped as suffering from "severe desertification" include not only the usual northwestern region but also Beijing, Tianjin, the moist and subtropical Sichuan basin, the southern double-cropped rice area, and the southwestern rice area. Dregne's estimates imply that the world's agriculture has lost between 8.7 percent and 30.1 percent of its total yields to desertification alone, though he does not say whether the desertification is per annum, per decade, or cumulated over all time. Sutton's subsequent USDA study reproduced Dregne's estimates to illustrate the environmental degradation associated with agriculture (Sutton 1989).

7. Wen notes that in the case of the loess plateau the share of soil lost from cultivated lands (30 percent) was higher than the share of all plateau lands that were cultivated (25 percent). This does indeed look like a case of farmers' "mining the soil." Soil mining is especially to be expected in the loess plateau, however, since the loess layer itself is still about 100 meters thick (Wen 1993, p. 65), thanks to millennia of dust buildup from upwind deserts. And even in this case, any summary of land losses should have deducted those from uncultivated areas if there is no separate evidence that human processes such as clear-cutting of forests are the source of the erosion from uncultivated lands.

8. Diemont, Smiet, and Nurdin (1991) made these same criticisms of the river siltation evidence.

9. Thorp 1939, p. 15. On the soil ecology of the loess plateau, see Institute of Soil Science, Academia Sinica (1990, chap. 6).

10. There is similar excess in the Post's warnings that the elevated Huang He will flood Kaifeng. One must pick one's threats carefully. As for the flooding of

the lower Huang He, some care must go into deciding whether to ring the flood alarm or to ring the no-water alarm. Unlike the *Post,* Lester Brown and others argue that the main problem between Kaifeng and the sea is the *absence* of water in the Huang He:

And the Huang He (Yellow River) in China, which flows through eight provinces, has run dry for part of each of the last eight years. As the provinces upstream divert more and more water for industrial and residential uses, the dry period grows longer. For several weeks in 1996, the river ran dry before reaching Shandong Province, the last one it flows through to the sea. In 1997, it went dry a week earlier than the preceding year. Unfortunately, the farmers in Shandong Province depend on the Huang He directly for underground water as it helps recharge the province's aquifers. (Brown *et al.* 1998, p. 7)

It is possible, with an extreme variation in weather conditions across the seasons, that the Huang He is a threat both because it increasingly runs dry and because it threatens to flood Kaifeng and other cities behind those high levies? But it is a low-probability combination of events. To clarify the real dangers, future discussions of this issue should trace out the likely dynamics more carefully than the literature on erosion has done.

11. Chapter 1 gave one example of this in Vaclav Smil's (1984) warnings about huge land losses in China since 1957. In a similar vein, Forestier (1989) warned of desertification so great that by A.D. 2000 there may be roughly twice the northern desert area as there was in 1949, despite China's courageous efforts to build a great green wall.

12. Tyler 1994, Wehrfritz 1995, p. 11; Pomfret 1998; and Brown et al. 1998, pp. 80–81, citing "Food Before Golf on Southern Land," *China Daily,* 25 January 1995. Going beyond golf courses alone, *Newsweek*'s Wehrfritz (1995, p. 11) worried about the effects of several dozen theme parks, several thousand graves, and 3,000 new industrial zones on China's cultivated area (which is about 142 million hectares). Brown (1995, chap. 4) expanded the list of grain land invaders to include vegetable plots, highways, shopping centers, tennis courts, and private villas. Chapter 6 quantifies the net impact of all the nonagricultural construction activity.

13. In 1990 and 12,800 golf facilities in the United States occupied an estimated 532,700 hectares (Balogh and Walker 1992, p. 8).

14. For an appreciation of the long-term field research tradition, see Jenkinson 1991, Mitchell et al. 1991, and the rest of the special January–February 1991 symposium issue of *Agronomy Journal.*

15. American Society of Agricultural Engineers 1985, pp. 40–47; and Lal and Pierce 1991. See also Pimentel 1993, chap. 12; Cruz, Francisco, and Conway 1988; Lal 1987; Van doren and Bartelli 1956; Buntley and Bell 1976.

16. Not all contemporary studies of soil that lack history deserve the criticisms that follow here. In particular, the highest scientific standard has been maintained by studies emanating from the Institute of Soil Science, Academia Sinica (e.g.,

1990) or from Indonesia's Centre for Soil and Agroclimate Research. Although their works are used below, they do not deal specifically with documenting longer-run trends in soil conditions.

17. Chapter 7 notes another study that usefully compares soil samples separated by more than a decade, in this case from the island of Sumatra in Indonesia. Van Noordwijk et al. (1997) matched soil profiles from the 1930s and 1990 in Lampung province to determine and explain the changes in organic carbon. Their database stays closer to the soil chemistry itself, however, incorporating a narrower range of nonsoil data than the Machakos District study or the current study.

Chapter 3

1. Two early drafts of such soil reference profiles report on our two countries: Institute of Soil Science, Academia Sinica and International Soil Reference and Information Centre (1994) for China, and Center for Soil and Agroclimate Research and International Soil Reference and Information Centre (1994).

2. The complete soil history data sets used in this book, along with the soil and agriculture data sets used in chapters 5, 6, and 8, are available from the author (Agricultural History Center, University of California–Davis, One Shields Avenue, Davis, CA 95616, USA) on a single compact disk, with accompanying explanations. A separate diskette file for the soil profiles of Hubei Province (China), which was inadvertently omitted from the compact disk, can be e-mailed as a separate Excel file attachment or mailed on a diskette. All these data sets are also available on-line at <http://aghistory.ucdavis.edu>.

3. Kenneth Pomeranz of the University of California–Irvine reports that there may be soil samples from the first decade of the twentieth century, taken by the British-American Tobacco Company in its search for new tobacco supply zones in China, at the Colonial Record Office in London, but they have not been pursued here.

4. The 1932–1944 profiles were published mainly in the *Soil Bulletins* of that era (China, 1936–1944, vols. 6–24), with an overlapping smaller set reproduced in Thorp 1939. Copies of the wartime *Bulletins* 21–24 were kindly supplied by the U.S. National Agricultural Library. As noted in note 2, the entire set of soil profile data used in this study is available from the author in a limited supply on compact disk.

For the story of prewar cooperation between Chinese and American soil scientists on soil surveys, the English language literature contains little beyond the description in Thorp 1939 and the *Soil Bulletins* themselves. Yet Chinese experts and American missionaries were already discussing plans for soil surveys as far back at 1903 (Stross 1986, pp. 18, 19, 85).

5. Professor Chen Ziming kindly made a copy of the first National Soil Survey available, and Wu Wanli discovered and utilized the Inner Mongolia study.

6. See the section of the references giving sources in Chinese.

7. An example of the intellectual continuity of China's soil science establishment in Nanjing is the long and distinguished career of Professor Li Chingkwei. Professor Chingkwei generated some of our primary data as a young field tester on the team involving Professor James Thorp in the 1930s, then was a leading organizer of the national soil surveys of 1959–1961 and 1981–1986. He also kindly offered advice on our project in 1993–1994.

8. On the techniques of measurement I am particularly grateful for conversations with Professor Li Chingkwei, who took many of the measurements in the 1930s. Among other things, he advised against using 1930s measurements of humus, of which several are included in our data set but excluded from the analysis. See also his description of the early measurement techniques for pH and organic matter in Thorp 1939, pp. 535–36.

9. These counts differ in a few places from those implied for the whole nation in table 3.1, because the present counts relate to statistical samples that were trimmed for want of some other variables in a few cases. Hence there are only thirty-one profiles for nitrogen for China in the 1930s, for example, even though table 3.1 implies thirty-five.

Chapter 4

1. For a more detailed comparison of average precipitation and temperature, see either Institute of Soil Science 1986 or the statistical yearbook of China.

2. There is a similar spatial correlation between precipitation and either temperature or altitude for Indonesia, forcing us to use precipitation alone as the main climate variable in our later discussion of Indonesia as well.

3. For these GLS equations the equation actually run is not the original OLS equation $Y = a + bH + cX + dS + e$, introduced in chapter 3, but rather the transformed equation $(Y/r) = a/r + b(H/r) + c(X/R) + d(S/r) + u$, which allows us to read the results as the original equation (again $Y = a + bH + cX + dS + e$), with the proviso that the error term $e = ru$ and r vary systematically with the data source.

4. For tobit equations on alkalinity in the north, see appendix table A.3.

5. Appendix B sketches the regression-implied geography of soil patterns in 1981–1986, by province. Its patterns may be compared with those in the *Soil Atlas* based on a slightly earlier set of data (Institute for Soil Science 1986).

6. The estimated averages of table 4.5 and figure 4.4 involve "noisy" back-casts from the 1980s back to the 1950s and 1930s in cases where the earlier era (either 1930s or 1950s) was based on six or more representative profiles for a particular region. To make the earlier periods imitate the mixture of soil types prevailing in the 1980s, we subtracted from each regional 1980s average for tilled fields only the change predicted by the historical shift terms and the change in the average prediction error for the profiles in that region (the change in "noise"). With Δ representing net changes from the earlier period to the 1980s, and the

underscored terms representing averages, the formula for each of table 4.5's numbers for the 1930s or 1950s is $\underline{Y}_{1980s} - b\Delta H - r\Delta \underline{u}$, which is equivalent to $\underline{Y}_{earlier} + c\Delta \underline{X} + d\Delta \underline{S}$ for the region's mixture of soils sampled in the 1980s.

7. For the chemistry of soil reaction, pH, salinity, and crop choice, see Institute of Soil Science 1990, chap. 27, Brady 1990, chap. 8, and the sources cited in each.

8. The full alkalinity regression results are relegated to appendix table A.2, which applies the less familiar tobit regression procedure appropriate to such a truncated dependent variable.

9. For this result, see the coefficients on the "Thorp map" variable in the Huang-Huai-Hai plain regressions in appendix tables A.2 and A.3. This variable equals one for any soil data where the pH figure was omitted from the profile data and had to be borrowed from the map on which figure 4.5 is based. Where the map was used, the pH value is the midrange value, or 9.0 for the top pH range.

10. In any case, the desert areas in question would constitute a very small share of north China's tilled lands.

11. For the full regressions on both acidity and pH, see appendix table A.5.

12. These trends, like all the trends mentioned in this chapter, are posited to have occurred on tilled fields. Since we expect no trend in the chemistry of untilled soils, it should be possible to show that each of the trends mentioned here was more evident on tilled than on untilled soils. We explored this expectation in other statistical tests that included an interaction between tillage and historic time. In each case, the trend tilt for tilled soils had the same sign as the trend described in the text, supporting our interpretation of trends as a sign of human influence on soil chemistry. None of the tillage and era interactions was statistically significant, however.

13. Erle Ellis and Wang Siming (1997) have given good historical and agronomic evidence suggesting that in the Lake Tai region just west of Shanghai, at least, total nitrogen has either held constant or risen for centuries, thanks to investments in organic manures and other conservationist practices.

14. The sole regional exception to this pattern is the southwestern rice region's 1950s–1980s improvement in organic matter, nitrogen, and potassium.

15. Exploring the behavior of bias clue measurements across regions and eras, we found only three cases in which they suggested a trend bias for north China:

1. For organic matter, investigators in the Shaanxi-Shanxi winter wheat region chose to study soil classes less rich in organic matter in both the 1930s and the 1950s. If there was a similar bias within soil types, this region, like the rest of the north, had a serious drop in organic matter from the 1930s to the 1950s.

2. Relative to the 1980s sample, the 1950s soils were drawn from classes that would have higher phosphorus and potassium levels in any time period. Any downward adjustment for this bias would imply that more of the rise in phosphorus and potassium came after the 1950s.

3. Relative to the 1980s sample, both the 1930s and the 1950s samples chose to study more alkaline types of soils in the spring wheat region and less alkaline types of soils in the winter wheat–millet region. If there was a similar bias hidden within each soil type, then adjusting for it would cancel figure 4.4's suggestion of any trends in pH in both regions.

For south China, the same test also suggested three minor biases:

4. For organic matter and for nitrogen, investigators seem to have chosen the 1930s and 1950s samples from the north Chang Jiang plain, mainly from Jiangsu, pessimistically (artificially lowering the average organic matter). Correcting for such a bias would steepen the slight decline in organic matter for this region, making it resemble the results for the neighboring Huang-Huai-Hai plain to the north.

5. For phosphorus and potassium, the 1950s samples seem to have chosen an optimistic mix of soils in the far south (double-cropping rice and southwestern rice regions). Correcting for this would accentuate the gain in phosphorus and potassium from the 1950s to the 1980s.

6. For pH, the 1930s and 1950s samples chose pessimistically in the tea-rice region, implying that the 1950s–1980s decline in pH should have been more severe. On the other hand, the 1930s and 1950s samples for the southwest rice region chose optimistically, suggesting that there may have been no reversion toward acidity in this region after the 1950s.

For details on the side tests, see Lindert, Lu, and Wu 1996a, pp. 1174–1176, and 1996b, pp. 338–339.

16. The source is Bruce Stone (1986, p. 470). The figures on fertilizer application show similar trends to these figures on production.

17. To pursue the possible fertilizer contributions further, one can follow two clues. First, the general decline in pH during the 1957 to 1984 period is probably related to the dramatic increases in the use of ammonium- and urea-type fertilizers in China during that time. Second, one can pursue the nutrient contribution ratios for China's organic fertilizers suggested by Wen (1984), who reckons that China's application of organic fertilizers in the rice zone doubled between 1952 and 1979.

Chapter 5

1. Miranowski and Hammes (1984), Gardner and Barrows (1985), and Peterson (1986) have used land prices to weigh the effects of soil quality and conservation in the United States. The procedure involved requires the existence of a well-integrated land market, as in the United States.

2. For a sampling of how authors as far back as the nineteenth century have noted and reacted to the human control over soil quality, see Hayami and Ruttan 1985, pp. 45–50.

3. Good examples of the production function approach that incorporate indirect measures of soil quality are Peterson 1986, Bhalla 1988, and Huang and Rozelle 1995.

4. For the full list of independent variables in this and other equations, see appendix table A.5. On fertilizer, irrigation, and technology in general, see Stone 1986, 1988.

5. In equations (2) and (3), when the dependent variable represents only crop yields and not total agricultural output, it is necessary to adjust all input measures to approximate inputs into those crops alone, not into all agriculture. This is not easy, in the absence of data dividing labor time, or irrigated area, or electricity among activities. Here we simplify, just multiplying the log of each input by the share of the relevant crops in the county's total value of agricultural product.

6. Specifically, the underlying lobbying model of fertilizer supply and of multiple cropping at the county (*xian*) level says that fertilizer supply and multiple cropping depend politically on the county's share of the province's urban population, collective-enterprise production, off-farm purchases, cotton production, and oil production, and the province's shares of the same at the national level. The full equations for fertilizer inputs and for multiple cropping are given in Lindert 1996, appendix F.

7. Institute of Soil Science 1986; Brady 1990.

8. Although we will use two-stage least squares, our specification is not fully conventional. The equations for acidity in the south and for alkalinity in the north are two-stage tobit. The bounded (nonnegative) nature of acidity and alkalinity as relevant to plant growth dictates the use of tobit. Yet the choice to use tobit mattered little to the effects of agriculture on the soil since unreported regressions using unbounded pH as the dependent variable gave similar results in both north and south China.

9. The agricultural data come primarily from the annual agricultural yearbooks, supplemented by employment data from the 1982 population atlas of China. I thank G. William Skinner for supplying a computerized version of the 1985 agricultural data, and Bing Zhang and Joann Lu for help in purging errors from the original data.

10. The estimated elasticities are evaluated at mean values of all independent variables. The elasticities vary over the sample range, because the equations are not fully log-linear.

11. A puzzle relating to organic matter and nitrogen is why their coefficients should have opposite signs for total GVAO in the south in equation (1). This appears to be an omitted-variable effect, in which lands with high organic matter to nitrogen ratios happen to be more productive for secondary crops in some way not captured by all the physical soil class variables. The most likely culprit is the incomplete reporting of land use in some of the southern provinces, where our interpretation of the raw data may have mistaken nitrogen-rich long-fallow lands for cultivated lands.

12. The favorable effect of non-main-crop yields on organic matter and nitrogen may be a hidden effect of organic fertilizers. Manure, green waste, and other organic fertilizers are not measured in the official fertilizer statistics and tend to be applied more intensively to specialty crops than to grains, oils, and cotton. On China's application of organic fertilizers since the 1930s, see Wen Qi-xiao 1984, pp. 45–56, and Stone 1986.

In addition, the non-main-crop category includes more nitrogen-fixing legumes and grasses.

Chapter 6

1. For the full spreadsheet journey, see Lindert 1996, appendix Q.

2. It may seem odd that a 1.0 percent change in the north and a 19.5 percent change in the south average out to only a 7.4 percent change for the nation, given that the land area in the south is greater. Yet the four regions included here in "south China" exclude the large Sichuan basin, for want of sufficient data from that region in the 1950s.

3. Here we implicitly assume that the ratio of the quality of unreported lands to the quality of reported lands is similar across provinces. That might or might not be the case. That is, the soil samples might or might not have been taken proportionately from both reported and unreported areas.

4. One anonymous reviewer doubts that reporting of cultivated area was nearly total in 1957 and counterproposes these rates of underreporting: "I would estimate underreporting in 1957 as 10 percent, 1930s as 25 percent, and 1978 as 25 percent as vague national percentages with an error on my estimates of about 20 percent for each period."

5. Let us use the popular imagery of "urbanization" here as a shorthand to represent the totality of all land conversions for construction of nonagricultural, nonforest, nonwilderness facilities: industrial and commercial sites, residences, and transportation infrastructure. Most, but not all, of these land uses are urban.

6. This assessment is based on Lindert, Lu, and Wu 1995, appendix U, the results of which were described in Lindert, Lu, and Wu 1996a, p. 1177.

7. Here we set aside trends in Xinjiang and Qinghai and Tibet for want of sufficient data on soil quality or on adjusted cultivated area. Adding these trends in, however, would reinforce the trends noted in the text. As table 6.3 shows, even the increasingly understated official figures on cultivated area show increases for these provinces.

8. See the regression results "with correction for rounding" in table 4.7, panel B.

Chapter 7

1. In what follows the shorthand "Java" is used as a shorthand for "Java and Madura."

2. The term "outer islands" is used as the best shorthand for all the islands of Indonesia other than Java and Madura. It includes Sumatra, Kalimantan, Sulawesi, Bali, Nusa Tenggara, Maluku, and Irian Jaya. Although being called "outer" may offend some sensibilities similar to those offended by the term "Far East," the geographical content of "outer" seems preferable to the usual alternative "other islands," which in isolation raises the distracting question "other than what"?

The outer islands are, of course, diverse in their climate, geology, and settlement history. For data convenience Bali is included among them despite its longer and denser settlement and its proximity to Java. An advantage of including Bali in the outer islands will be reaped in the next chapter, when it helps provide statistical diversity in the outer-island samples relating soil to agricultural productivity.

3. Having data on soils subject to shifting cultivation is attractive for property rights economists wanting to test for any effects of this loose land right regime on humans' treatment of the soil. Unfortunately for that purpose, soils mentioning shifting cultivation are less than 1 percent of the Java-Madura historical soil sample and only about 3 percent of the outer-island sample. Heavier data collection, as in the government's agricultural data for the whole country in 1990, show great shares of shifting cultivation in total arable land area, though even these shares are limited: under 3 percent for Java-Madura and about 10 percent for the outer islands.

4. A fuller specification of the soil site's climate would involve altitude as well as precipitation, since altitude affects both precipitation and temperature. Yet the site altitude is seldom given in the underlying primary data. Knowing the village, and occasionally something about the site's physical setting, makes it easier to locate a particular site in terms of the CSAR's map of precipitation zones than to guess at its altitude, which can vary abruptly on the volcanic islands. In any case, the land use variable often captures the effect of altitude, for example, when we are told that the site is paddy or *tegalan* or an upland tree crop.

5. A wet month is defined for the underlying CSAR map as a month with average precipitation of more than 200 millimeters, and a dry month as one averaging less than 100 millimeters.

For data convenience, figure 7.2 maps mainly the district precipitation averages to be used in chapter 8's statistical analysis. Where the district division is seriously misleading about the climate geography, as in Kalimantan and Irian Jaya, freehand zone divisions have been supplied, based on the original unpublished CSAR map.

6. The spline function imitates the time path of a polynomial time function, with the advantage that the discrete time slopes used here make the trends more transparent than would a polynomial of order four or higher.

7. Although this set of time and land use interactions takes advantage of the chance to test how different land uses change soils over time, it is not the only set of interactions that researchers may wish to try on such a data set. For

example, one might think of physical processes in which the time trend might vary with soil class for given land uses. As least one recent study on Tanzanian data found that trends since the 1950s vary among (five) soil classes even with the same history of land use (Hartemink 1997). I have not explored this possibility here. None of the five soil classes from the Tanzanian study was found in Indonesia, preventing any useful comparison. More generally, the task of searching over all trend/soil class combinations would be a larger one. The Indonesian data set involves at least fourteen soil classes in any one regression. Mapping the interactions of each of these with six time slope variables already creates eighty-four new variables to sift through, even without considering interactions with parent material, climate, and so forth. This study has therefore been confined to an additive specification of the roles of all the physical variables.

8. For support for the envisioned dynamic for organic matter, nitrogen, and the ratio of the two, see the forest to field to commercial tree dynamic documented for Lampung by Van Noordwijk et al. (1997).

9. One should bear in mind that the choice of 1970 as an approximate switch point for some of Indonesia's soil trends rests on the statistical handling of the anomaly of lower average nitrogen, phosphorus, and potassium for the period 1974–1986. In this chapter, I tentatively interpret the lower nitrogen, phosphorus, and potassium levels of these years as a false result caused by changes in either laboratory measurement or soil-sampling selectivity.

Yet the possibility that something actually lowered levels of these soil nutrients in the 1974–1986 period cannot be ruled out. In Indonesia, more than in most countries, the decade ending in 1980 saw a particular kind of agricultural revolution, one that could have transformed its soil chemistry. Output suddenly jumped after 1970, or more accurately after 1967. As late as 1967, Java's agricultural output still had not surpassed its level of 1940, despite a 49 percent growth in population. From 1967 on, it grew 42–43 percent per decade. Perhaps one main difference, then, between the 1970s and the preceding three decades was the sudden strong uptake of available nutrients from the soil by crops and related plant matter.

The agricultural intensification of the 1970s also brought what one might call the "urea boom." Urea dominated the surge in fertilizer use from about 1974 to 1983, helped by rising subsidies (Booth 1988, chap. 5; Adiningsih, Santoso, and Sudjadi 1990). Urea is capable of raising nitrogen availability to plants very rapidly, with possibly favorable effects on the soil reserves of both total nitrogen and organic matter. Its heavy ammonia reactions tend to lower soil pH. The urea boom may have lowered pH from about 1970 to the early 1980s, before the fertilizer policy shifted toward triple super phosphate and other fertilizers. The reliance on urea, plus the heavy uptake from rising yields, may also have lowered soil phosphorus and potassium. Studies have found that applying nitrogen fertilizers triggers high removal rates for potassium in particular (IRRI-Cornell 1980, pp. 265–67). Thus the sudden increase in urea applications, followed by a shift toward fertilizers supplying more phosphorus and potassium, could help explain a true dip then rise in levels of those two nutrients.

A sudden rise in fertilizer application inputs, however, particularly of urea, does not square with a drop in soil nitrogen by a third or more around 1980, as the uncorrected raw data imply. My interpretation is that the nitrogen-phosphorus-potassium dip in 1974–1986 was a mirage, and that I have adjusted the relevant data correctly with the regression-based trends shown in figure 7.3 and table 7.4.

10. Again, acidity is simply expressed as the pH below 6.

11. One should not infer that Java had a direct advantage in climate, however, for any given state of the soil. Chapter 8's exploration of the determinants of agricultural productivity on the different islands in 1990 implies that the yield-maximizing pattern of precipitation in Indonesia tended to be somewhat wetter than the pattern in Java's precipitation. Therefore, if climate favored Java, it would have to have conveyed its advantage in the process of soil genesis.

Chapter 8

1. An attractive direct approach might be to sort out the whole simultaneous agriculture → soil and soil → agriculture system by using historical data alone. One could view every single year's agricultural conditions as a separate agricultural environment and run simultaneous-equation tests over years instead of over space. Data constraints rule out this attractive alternative, however. We cannot get good data, even national aggregates, on all the relevant variables for Indonesia for years before 1990.

2. The predictions from the historical soil sample eliminate random noise, sample depth, and pre-1990 human impacts on the soil. First, each historical profile is converted from an observed value to a predicted value (e.g., from observed $\ln(P_2O_5)$ to the predicted $\ln(P_2O_5)$) by subtracting the error term from the observed value. Then the regression coefficient relating to soil depth is used to adjust the predicted value to what it would be if the topsoil sample extended from the surface down to 10 centimeters. Finally, the historical value (e.g., a value of $\ln(P_2O_5)$ in 1957) is adjusted to a value predicted for the year 1990 using the time terms of chapter 7's spline function.

Only those districts are included here for which there were profiles in the main land use categories, as described in chapter 7. For these the average chosen is, to repeat, not just the observed average of those few soils from that district, but the smoother average of the values *predicted* for those few soils.

This chapter's agricultural samples exclude DKI Jakarta and East Timor.

3. The production function for Approach A, as shown in equations (1)–(3) of table 8.1, is a modified log-log Cobb-Douglas function, which is a special case of chapter 5's translog function. The simpler special case is used here for two reasons. First, chapter 5's experience suggested that adding the extra terms of the higher-order translog function added little predictive power, indicating that the extra terms are not worth the burden of discussing them. Second, the Indonesian samples are smaller: 77 districts in Java and 96 in the outer islands, versus 161 counties of north China and 282 counties of south China.

4. Those input variables that take on zero values for some districts could not be put in logarithmic form: machinery, fertilizer, and soil acidity. Their effects are judged from a semilog form here, relating the log of agricultural output per harvested hectare to the absolute level of these inputs per hectare.

Another simplification is that certain variables were omitted because they were so highly correlated with those that were included that multicollinearity would have confounded estimating their effect reliably. Thus the reported phosphorus and potassium contents of applied fertilizer were omitted in favor of the included value of the nitrogen content of applied fertilizers. Soil nitrogen was omitted because it is highly correlated with soil organic matter, as one would expect.

5. For fertilizer, which is applied much more heavily on Java than elsewhere, there is also the likelihood that the marginal product of extra nitrogen-phosphorus-potassium content is near zero. Soil scientists have shown that for Indonesia, as textbooks report for the whole world as well, the gains from extra nitrogen or phosphorus or potassium drop close to zero at certain thresholds of availability, thresholds that may have been reached in Java before 1990. For phosphates in particular, see Amar 1987 and Diamond et al. 1988.

6. The ratio of arable area to harvested area in 1990 was 1.82 for the outer islands but only 0.75 for Java. If we interpret the arable area in Java as an approximation of the true area cultivated any time in a year, then the implied average MCI is 1/0.75 = 1.33 crops each year in Java. On the other islands, a ratio as high as 1.82 should not be interpreted as only 1/1.82 = 0.55 harvests per year on the average cultivated hectare, but simply an abundance of lands reported as arable that are not cultivated at all in any given year.

7. For "organic matter" here read "organic matter and total nitrogen." As mentioned in note 4, the two properties are highly correlated across soils and tend to cycle around relatively constant ratios in the absence of shocks. Their high correlation means that one of the two, here organic matter, must be chosen to represent both to avoid multicollinearity.

8. Indonesian experts have recently improved this substitution, by showing how to increase plant uptake of nitrogen without loss of soil organic matter or nitrogen through deep planting of urea (Ismunadji 1990).

Nitrogen experiments in the Philippines by IRRI are also relevant here. Cassman et al. 1995 and Cassman et al. 1996 report experimental results showing that the nitrate fertilizer has disappearing ability to promote plant uptake and sustain soil N levels in fields where soil N levels are high. This suggests substitution of fertilizer for soil endowment mainly at lower soil N levels, as in Java, a result consistent with the spatial pattern of the current regression results.

9. For any year t before 1990,

$$\text{soil chemical quality index} = 100 \cdot \log^{-1}\left(\sum_i (b_i \cdot \ln(x_{it}/x_{i,1990}))\right),$$

where b is a coefficient from table 8.1, X is the level of a given soil chemistry characteristic, t is this year, and i is each soil characteristic (organic matter, P_2O_5, or K_2O) having a significant productivity effect.

10. It is better to use the narrower term "soil chemical quality" instead of the preferred term "soil quality" because data limitations prevented any detailed quantification of all the separate soil characteristics (e.g., texture, water table, gradient) that may have productivity consequences beyond their effect on soil organic matter, nitrogen, phosphorus, potassium, and pH. Even some chemical features, such as trace minerals, must be omitted here.

Chapter 9

1. Long-run experiments by the International Rice Resource Institute, backed by data from nonexperimental farms as well, suggest that for the lowland rice zone there is no easy translation from extra chemical fertilizer applications to yields beyond those rates of application already practiced. In this zone, which covers much of south China and Indonesia, there is even a poor correlation between total soil organic matter and total nitrogen on the one hand and the plant-available nitrogen supply on the other (Cassman et al. 1993, Cassman et al. 1995, and Cassman et al. 1996).

2. This basic point about China's comparative advantage and its implications for China's true food security has been made before. For recent statements, see Johnson 1998 and Lindert and Pugel 1996, pp. 48–49, 293–95.

3. To say that under the best of policies China could wisely replace grain lands with all these other uses does not, of course, imply that *any* policy replacing grain lands with these other uses is necessarily a wise one.

4. It was not possible to conduct a similar test for Indonesia in chapter 7 or 8. To see why, recall the approach to estimating the agriculture → soil characteristics linkage in the Indonesian case, where labor and other inputs were free to migrate between districts. In that case, we had to make all inferences of the effects of agriculture on the soil from the historical time dimension, not from the kind of cross-sectional simultaneous system estimated for China in chapter 5. In the historical time dimension, we lacked pre-1990 data on the agricultural product mix by district (*kabupaten*).

References

I. General

Ahmed, Yusuf, J., Salah El Serafy, and Ernst Lutz, eds. 1989. *Environmental Accounting for Sustainable Development*. A UNEP-World Bank Symposium. Washington, D.C.: The World Bank.

American Society of Agricultural Engineers. 1985. *Erosion and Soil Productivity*. Proceedings of the National Symposium on Erosion and Soil Productivity, New Orleans, Louisiana. St. Joseph, Mich.: Author.

Balogh, James C., and William J. Walker. 1992. *Golf Course Management and Construction: Environmental Issues*. Boca Raton, Fla.: Lewis Publishers.

Barbier, Edward B. 1996. "The Economics of Soil Erosion: Theory, Methodology, and Examples." University of York, Discussion Papers in Environmental Economics and Environmental Management, no. 9602, April.

Barbier, Edward B., and Joshua T. Bishop. 1995. "Economic Values and Incentives Affecting Soil and Water Conservation in Developing Countries." *Journal of Soil and Water Conservation* 50 (March–April): 133–37.

Bhalla, Surjit A. 1988. "Does Land Quality Matter? Theory and Measurement." *Journal of Development Economics* 29: 45–62.

Brady, Nyle C. 1990. *The Nature and Properties of Soils*. 10th ed. New York: Macmillan.

Brown, Lester R., and E. C. Wolf. 1984. "Soil Erosion: Quiet Crisis in the World Economy." Worldwatch Paper 60, Worldwatch Institute, Washington, D.C.

Brown, Lester R., Christopher Flavin, and Hilary French. 1998. *State of the World 1998*. New York: W.W. Norton.

Buntley, G. J., and F. F. Bell. 1976. *Yield Estimates for Major Crops Growth on Soils of West Tennessee*. Bulletin no. 561. Knoxville: Tennessee Agricultural Experiment Station.

Cassman, K. G., S. K. De Datta, D. C. Olk, J. Alcantara, M. Samson, J. Descalsota, and M. Dizon. 1993. "Yield Decline and the Nitrogen Economy of Long-

Term Experiments on Continuous, Irrigated Rice Systems in the Tropics." In Rattan Lal and B. A. Stewart (eds.), *Soil Management: Experimental Basis for Sustainability and Environmental Quality.* Boca Raton, Florida: Lewis/CRC Publishers, pp. 181–222.

Cassman, K. G., S. K. De Datta, D. C. Olk, J. Alcantara, M. Samson, J. Descalsota, and M. Dizon. 1995. "Yield Decline and the Nitrogen Economy of Long-Term Experiments on Continuous, Irrigated Rice Systems in the Tropics." in R. Lal and B. A. Stewart, eds., *Soil Management: Experimental Basis for Sustainability and Environmental Quality,* 181–222. Boca Raton, Fla.: Lewis/CRC Publishers.

Cassman, K. G., A. Dobermann, P. C. Santa Cruz, G. C. Hines, M. I. Samson, J. P. Descalsota, J. M. Alcantara, M. Dizon, and D. C. Olk. 1996. "Soil Organic Matter and the Indigenous Nitrogen Supply of Intensive Irrigated Rice Systems in the Tropics." *Plant and Soil* 182: 267–78.

Crosson, Pierre. 1985. "National Costs of Erosion Effects on Productivity." In American Society of Agricultural Engineers, *Erosion and Soil Productivity* (Proceedings of the National Symposium on Erosion and Soil Productivity, New Orleans, Louisiana), 254–65. St. Joseph, Mich.: American Society of Agricultural Engineers.

Crosson, Pierre. 1995. "Soil Erosion and Its On-Farm Productivity Consequences: What Do We Know?" Resources Discussion Paper 95–29, Resources for the Future, Washington, DC, June.

Crosson, Pierre, and Anthony Stout. 1983. *Productivity Effects of Cropland Erosion in the United States.* Washington, DC: Resources for the Future.

Cruz, W., H. A. Francisco, and Z. T. Conway. 1988. "The On-Site and Downstream Costs of Soil Erosion in the Magat and Pantabangan Watersheds." *Journal of Philippine Development,* 15(26): 85–111.

Doran, J. W., D. C. Coleman, D. F. Bexdicek, and B. A. Stewart. 1994. *Defining Soil Quality for a Sustainable Environment.* Madison, Wis.: Soil Science Society of America and American Society of Agronomy.

Dregne, Harold E. 1982. *Impact of Land Degradation on Future World Food Production.* Washington, D.C.: International Economics Division, Economic Research Service, U.S. Department of Agriculture.

Dregne, Harold E. 1992. "Erosion and Soil Productivity in Asia." *Journal of Soil and Water Conservation* 47 (January–February): 8–13.

Eckholm, Erik P. 1976. *Losing Ground: Environmental Stress and World Food Prospects.* New York: W.W. Norton.

Ellis, E. C., and Wang Siming. 1997. "Sustainable Traditional Agriculture in the Tai Lake Region of China." *Agriculture, Ecosystems & Environment* 61: 177–93.

El-Swaify, S. A., W. C. Moldenhauer, and Andrew Lo, eds. 1985. *Soil Erosion and Conservation.* Ankeny, Iowa: Soil Conservation Society of America.

Follett, R. F., and B. A. Stewart, eds. 1985. *Soil Erosion and Crop Productivity.* Madison, Wis.: American Society of Agronomy, Crop Science Society of America, and Soil Science Society of America.

Food and Agriculture Organization. 1979. *A Provisional Methodology for Soil Degradation Assessment.* Rome: Food and Agriculture Organization of the United Nations.

Food and Agriculture Organization. 1995. *Production Yearbook 1995.* Rome: Food and Agriculture Organization of the United Nations.

Frye, W. W., Bennett, O. L., and Buntley, G. J. 1985. "Restoration of Crop Productivity on Eroded or Degraded Soils." In R. F. Follett and B. A. Stewart, eds., *Soil Erosion and Crop Productivity,* 339–57. Madison, Wis.: American Society of Agronomy.

Gardner, Gary. 1996. "Asia is Losing Ground." *World Watch* 9 (November–December): 19–27.

Gardner, K., and R. Barrows. 1985. "The Impact of Soil Conservation Investment on Land Prices." *American Journal of Agricultural Economics* 67(4): 943–47.

Harroy, Jean-Paul. 1949. *Afrique—Terre Qui Meurt.* Brussels: Marcel Hayez.

Hartemink, Alfred E. 1997. "Soil Fertility Decline in some Major Soil Groupings under Permanent Cropping in Tanga Region, Tanzania." *Geoderma* 75: 215–229.

Hayami, Yujiro, and Vernon W. Ruttan. 1985. *Agricultural Development: An International Perspective.* Rev. and exp. ed. Baltimore: Johns Hopkins University Press.

Indian Council of Agricultural Research. 1957. *Final Report of the All-India Soil Survey Scheme.* Calcutta: Government of India Press.

International Rice Research Institute and Cornell University (IRRI-Cornell). 1980. *Priorities for Alleviating Soil-Related Constraints to Food Production in the Tropics.* Los Baños, Philippines: Author.

Jenkinson, D. S. 1991. "The Rothamsted Long-Term Experiments: Are They Still of Use?" *Agronomy Journal* 83(1): 2–10.

Johnson, D. Gale. 1995. "China's Future Food Supply: Will China Starve the World?" University of Chicago, Chicago, Ill. Typescript.

Johnson, D. Gale. 1998. "The Growth of Demand Will Limit Output Growth for Food over the Next Quarter Century." Paper No. 98:10, Office of Agricultural Economics Research, University of Chicago, Chicago, Ill., August 22.

Judson, Sheldon. 1968. "Erosion of the Land, or What's Happening to Our Continents." *American Scientist* 56.

Lal, Rattan. 1987. "Effects of Erosion on Crop Productivity." *Critical Reviews in Plant Sciences* 5(4): 303–67.

Lal, Rattan. 1990. *Soil Erosion in the Tropics: Principles and Management.* New York: McGraw-Hill.

Lal, Rattan, and F. J. Pierce, eds. 1991. *Soil Management for Sustainability.* Ankeny, Iowa: Soil and Water Conservation Society.

Larson, W. E., F. J. Pierce, and R. H. Dowdy. 1983. "The Threat of Soil Erosion to Long-Term Crop Production." *Science* 219: 458–465.

Lindert, Peter H., and Thomas A. Pugel. 1996. *International Economics.* 10th ed. Chicago: Richard D. Irwin.

Maddala, G. S. 1983. *Limited Dependent and Qualitative Variables in Econometrics.* Cambridge: Cambridge University Press.

Miranowski, John A., and Brian D. Hammes. 1984. "Implicit Prices of Soil Characteristics for Farmland in Iowa." *American Journal of Agricultural Economics* 66: 745–49.

Mitchell, C. C., R. L. Westerman, J. R. Brown, and T. R. Peck. 1991. "Overview of Long-Term Agronomic Research." *Agronomy Journal* 83(1): 24–28.

Oldeman, L. R., R. T. A. Hakkeling, and W. G. Sombroek. 1991. *World Map of the Status of Human-Induced Soil Degradation: An Explanatory Note* (including the map itself). Rev. 2d ed. Wageningen, Netherlands: International Soil Reference and Information Center, United Nations Environment Programme.

Oldeman, L. R., V. W. P. van Engelen, and J. H. M. Pulles. 1990. "The Extent of Human-Induced Soil Degradation." In L. R. Oldeman, R. T. A. Hakkeling, and W. G. Sombroek, *World Map of the Status of Human-Induced Soil Degradation: An Explanatory Note,* rev. 2d ed., Annex 5. Wageningen, Netherlands: International Soil Reference and Information Centre, United Nations Environment Programme.

Peterson, Willis. 1986. "Land Quality and Prices." *American Journal of Agricultural Economics* 68(4): 812–19.

Pierce, F. J., W. E. Larson, R. H. Dowdy, and W. A. P. Graham. 1983. "Productivity of Soils—Assessing Long-Term Changes Due to Erosion." *Journal of Soil and Water Conservation* 38(1): 39–44.

Pimentel, David, ed. 1993. *World Soil Erosion and Conservation.* Cambridge: Cambridge University Press.

Raychaudhuri, S. P., R. R. Agarwal, N. R. Datta Biswas, S. P. Gupta, and P. K. Thomas. 1963. *Soils of India.* New Delhi: Indian Council of Agricultural Research.

Rijsbergen, Frank R., and M. Gordon Wolman, eds. 1984. *Quantification of the Effect of Erosion on Soil Productivity in an International Context.* Delft, Netherlands: Delft Hydraulics Laboratory.

Scherr, Sara J. 1999. "Soil Degradation: A Threat to Developing-Country Food Security by 2020?" Food, Agriculture, and the Environment Discussion Paper 27, International Food Policy Research Institute,

Simon, Julian L. 1981. *The Ultimate Resource.* Princeton: Princeton University Press.

Social Science Research Council, Committee on the Economy of China. 1969. *Provincial Agricultural Statistics for Communist China,* Ithaca, N.Y.: Author.

Stocking, Mike. 1987. "Measuring Land Degradation." In Piers Blaikie and Harold Brookfield, eds., *Land Degradation and Society,* 49–63. London: Methuen.

Sutton, John. 1989. "Environmental Degradation and Agriculture." In *World Agriculture: Situation and Outlook Report* (special issue). Washington, D.C.: U.S. Department of Agriculture, Economic Research Service.

Tiffen, Mary, Michael Mortimore, and Francis Gichuki. 1994. *More People, Less Erosion: Environmental Recovery in Kenya.* New York: J. Wiley.

Trimble, Stanley W. 1974. *Man-Induced Soil Erosion on the Southern Piedmont, 1700–1970.* Ankeny, Iowa: Soil Conservation Society of America.

United Nations Conference on Desertification. 1977. *Desertification: Its Causes and Consequences.* New York: Pergamon Press.

Van Doren, C. A., and L. J. Bartelli. 1956. "A Method of Forecasting Soil Loss." *Agricultural Engineering* 37: 335–41.

World Bank. 1993. *World Development Report 1993.* New York: Oxford University Press.

World Bank. 1997. *World Development Report 1997.* New York: Oxford University Press.

World Resources Institute in collaboration with the United Nations Environment Programme and the United Nations Development Programme. 1990. *World Resources, 1990–91.* New York: Oxford University Press. Editions for 1986–1988/89 in collaboration with the International Institute for Environment and Development, published by Basic Books, New York.

World Resources Institute in collaboration with the United Nations Environment Programme and the United Nations Development Programme. 1992. *World Resources, 1992–93.* New York: Oxford University Press.

World Resources Institute in collaboration with the United Nations Environment Programme and the United Nations Development Programme. 1994. *World Resources, 1994–95.* New York: Oxford University Press.

II. Sources in Chinese

Chifeng Shi (District), The Office of Soil Survey. 1989. *ChiFeng Shi TuRang* (The Soils of Chifeng Shi (District)). Huhehaote: Inner Mongolia People's Press. May.

China. Chung yang ti chih tiao ch'a so. *T'u Jang Chuan Pao* (Soil Bulletin). Volumes 1 (1930)–24 (1944).

China, Office of the National Soil Survey. 1993. *ZhongGuo TuZhong Zhi* [Manual of the Soil Species of China]. 2 vols. Beijing: Agricultural Press. April.

China, Office of the National Soil Survey. 1996. *ZhongGuo TuRang Pu Cha Shu Jiu* (China's Soil Census Statistics). Beijing: China Agricultural Department.

Deng Chenyi, Wu Wenchao, and Lu Zhenxing. 1985. *TuMoTe ZuoQi TuRang* (The Soils of Tumote Zuoqi). Huhehaote: Inner Mongolia People's Press. May.

Gansu Province, Office of Soil Survey. 1991. *GanSu TuRang* (The Soils of Gansu). First draft. April.

Guangdong Province, Office of Soil Survey. 1993. *GuangDong TuRang* (The Soils of Guangdong). Beijing: Science Press. April.

Guangxi Province, Office of Soil Survey. 1990. *GuangXi TuRang* (The Soils of Guangxi). Final draft. June.

Guizhou Province, Office of Soil Survey. 1993. *GuiZhou TuRang* (The Soils of Guizhou). Final draft.

Hebei Province, Office of Soil Survey. 1990. *HeBeiTuRang* (The Soils of Hebei). Shijiazhuang: Hebei Science and Technology Press. December.

Heilongjiang Province. *The Bureau of Land Management and the Office of Soil Survey, 1992. HeiLongJiang TuRang* (The Soils of Heilongjiang). Beijing: Agricultural Press.

Henan Province, Office of Soil Survey. 1991. *HeNan TuRang* (The Soils of Henan). In-press draft.

Hubei Province, Office of Soil Survey. 1992. *HuBei TuZhong Zhi* (Manual of the Soil Species of Hubei Province). In-press draft.

Hunan Province, Bureau of Agriculture, 1989. *HuNan TuRang* (The Soils of Hunan). Beijing: Agricultural Press. May.

JiaWenJin. 1992. *LiaoNing TuRang* (The Soils of LiaoNing). ShenYang: Liaoning Science and Technology Press. December.

Jilin Province, The Office of Soil Survey. 1991. *JiLin TuRang* (The Soils of Jilin). In-press draft.

NingXia Province, Institute of Agricultural Survey and Planning. 1990. *NingXia TuRang* (The Soils of Ningxia). Yinchuan: Ningxia People's Press. September.

Shaanxi Province, Office of Soil Survey. 1992. *ShaanXi TuRang* (The Soils of Shaanxi). Beijing: Science Press. August.

Shandong Province, Center of Soil Fertility. 1993. *ShanDong TuZhong Zhi* (Manual of the Soil Species of Shandong). Beijing: Agricultural Press. December.

Shanxi Province, Office of Soil Survey. 1992. *ShanXi TuRang* (The Soils of Shanxi). Beijing, Science Press. November.

Sichuan Province. 1994. Preliminary draft for volume on the soils of Sichuan.

The Team of Inner Mongolia, and NingXia Comprehensive Survey, and Institute of Soil Science in Nanjing, The Chinese Academy of Sciences. 1978. *NeiMengGuo ZiZhiQu Yu DongBei XiBu DiQu TuRang DiLi* (Soil Geography of Inner Mongolia Autonomous Region and West Region of North-East China). Beijing, Science Press. June.

Yunnan Province, Office of Soil Survey. 1993. *YunNan TuRang* (The Soils of Yunnan). Final draft.

III. China—In English and Russian

Brown, Lester R. 1995. *Who Will Feed China?* New York: W.W. Norton.

Buck, John Lossing. 1937. *Land Utilization in China: Statistics.* Chicago: University of Chicago Press.

China, State Statistical Bureau. 1996. *China Statistical Yearbook 1996.* Beijing: China Statistical Publishing House.

Chao, Kang. 1970. *Agricultural Production in Communist China, 1945–1965.* Madison: University of Wisconsin Press.

Crook, Frederick W. 1992. "Primary Issues in China's Grain Economy in the 1990s." In U.S. Congress, Joint Economic Committee, *China's Economic Dilemmas in the 1990s,* 385–402. Armonk, N.Y.: Sharpe.

Dawson, Owen L. 1970. *Communist China's Agriculture.* New York: Praeger.

"Malthus Goes East." 1995. *Economist,* August 12, 29.

Field, Robert Michael. 1993. "Trends in the Value of Agricultural Output, 1978–1988." In Y. Y. Kueh and Robert F. Ash, eds, *Economic Trends in Chinese Agriculture.* Oxford: Clarendon Press, pp. 123–160.

Forestier, K. 1989. "The Degreening of China." *New Scientist* 123(1671): 52–58.

Heilig, Gerhard K. 1997. "Anthropogenic Factors in Land-Use Change in China." *Population and Development Review* 23(1): 139–68.

Huang Jikun and Scott Rozelle. 1995. "Environmental Stress and Grain Yields in China." *American Journal of Agricultural Economics* 77 (November): 853–64.

Institute of Soil Science, Academia Sinica. 1986. *Chung-kuo t'u jang t'u chi.* (The Soil Atlas of China). Beijing: Cartographic Publishing House.

Institute of Soil Science, Academia Sinica. 1990. *Soils of China.* Beijing: Science Press.

Institute of Soil Science, Academia Sinica, and International Soil Reference and Information Centre. 1994. "Soil Reference Profiles of the People's Republic of China: Field and Analytical Data." Country Report 2, Draft, July.

Khan, Azizur Rahman, Keith Griffin, Carl Riskin, and Zhao Renwei. 1993. "Household Income and Its Distribution in China." In Keith Griffin and Zhao Renwei, eds., *The Distribution of Income in China.* New York: St. Martin's Press, pp. 25–73.

Kovda, Viktor Abramovich. 1959. *Ocherki Prirody i Pochv Kitaia* (Notes on the Nature and Soils of China.) Moscow: Izdatel'stvo Akademii Nauk SSSR.

Kueh, Y. Y., and Robert F. Ash, eds. 1993. *Economic Trends in Chinese Agriculture: The Impact of Post-Mao Reforms.* Oxford: Clarendon Press.

Lindert, Peter H. 1996. "Appendices on China Soils, 1930s—1980s." Working paper no. 84, Agricultural History Center, University of California–Davis, December.

Lindert, Peter H., Joann Lu, and Wu Wanli. 1995. "Soil Trends in China since the 1930s." Working paper no. 79, Agricultural History Center, University of California–Davis, August.

Lindert, Peter H., Joann Lu, and Wu Wanli. 1996a. "Trends in the Soil Chemistry of North China since the 1930s." *Journal of Environmental Quality* 25(4): 1168–78.

Lindert, Peter H., Joann Lu, and Wu Wanli. 1996b. "Trends in the Soil Chemistry of South China since the 1930s." *Soil Science* 161 (May): 329–42.

Marks, Robert B. 1998. *Tigers, Rice, Silk, and Silt: Environment and Economy in Late Imperial South China.* Cambridge: Cambridge University Press.

McLaughlin, L. 1993. "A Case Study in Dingxi County, Gansu Province, China." In David Pimentel, ed., *World Soil Erosion and Conservation,* 87–108. Cambridge: Cambridge University Press.

Orleans, Leo A. 1992. "Loss and Misuse of China's Cultivated Land." In U.S. Congress, Joint Economic Committee, eds., *China's Economic Dilemmas in the 1990s: The Problems of Reforms, Modernization, and Interdependence,* pp. 403–417. Armonk, N.Y.: M. E. Sharpe,

Parham, Walter E., Patricia J. Durana, and Alison L. Hess, eds. 1993. *Improving Degraded Lands: Promising Experiences from South China.* Honolulu: Bishop Museum Press.

Pomfret, John. 1998. "Decades of Misuse Turn Prairies to Dust—China Digs In to Reverse Costly Land Loss." *Washington Post,* November 9.

Pomeranz, Kenneth. 1993. *The Making of a Hinterland: State, Society, and Economy in Inland North China, 1853–1937.* Berkeley and Los Angeles: University of California Press.

Pomeranz, Kenneth. 2000. *Economy, Ecology, Comparisons, and Connections: The World in the Age of the Industrial Revolution.* Princeton: Princeton University Press.

Smil, Vaclav. 1984. *The Bad Earth: Environmental Degradation in China.* Armonk, N.Y.: M.E. Sharpe.

Smil, Vaclav. 1987. "Land Degradation in China: An Ancient Problem Getting Worse." In Piers M. Blaikslie and Harold Brookfield, eds., *Land Degradation and Society,* 214–22. London: Methuen.

Stone, Bruce. 1986. "Chinese Fertilizer Application in the 1980s and 1990s: Issues of Growth, Balance, Allocation, Efficiency and Response." In U.S. Congress, Joint Economic Committee, *China's Economy Looks toward the Year 2000,* 1:453–96. Washington: U.S. Government Printing Office.

Stone, Bruce. 1988. "Developments of Agricultural Technology." *China Quarterly* no. 116 (December): 767–822.

Stross, Randall E. 1986. *The Stubborn Earth: American Agriculturalists on Chinese Soil 1898–1937.* Berkeley and Los Angeles: University of California Press.

Thorp, James. 1939. *Geography of the Soils of China.* Peking: National Geological Survey of China.

Tyler, Patrick E. 1994. "Nature and Economic Boom Devouring China's Farmland." *New York Times,* March 24, International section.

van Lynden, G. W. J. 1998. "Assessment of the Status of Human-Induced Soil Degradation in China." Paper presented to the Working Meeting of the Strategy and Action Project for Chinese and Global Food Security, February 18–19, International Food Policy Research Institute, Washington, D.C.

Wang Tong et al. 1992. *An Economic Analysis of the Progressive Decrease in the Quantity of China's Cultivated Land.* Beijing: Jingji Kexue Chubanshe.

Wang Xiaoju and Gong Zitong. 1998. "Assessment and Analysis of Soil Quality Changes after Eleven Years of Reclamation in Subtropical China," *Geoderma* 81: 339–55.

Wehrfritz, George. 1995. "Grain Drain: China's Peasants Are Reducing Their Crops of Wheat and Rice. The Result Will Probably Be a Future Grain Shortage of Disastrous Dimensions." *Newsweek,* International Edition, May 15, 8–14.

Wen Dazhong. 1993. "Soil Erosion and Conservation in China." In David Pimentel, ed., *World Soil Erosion and Conservation,* 63–86. Cambridge: Cambridge University Press.

Wen Qi-xiao. 1984. "Utilization of Organic Materials in Rice Production in China." In *Organic Matter and Rice,* 45–56. Los Baños, Philippines: International Rice Research Institute.

Wu, C., and H. Guo, eds. 1994. *Land Use in China.* Beijing: Science Press.

IV. Indonesia

Adiningsih, J. Sri, Djoko Santoso, and M. Sudjadi. 1990. "The Status of N, P, K, and S of Lowland Rice Soils in Java." In Graeme Blair and Rod LeFloy, eds., *Sulfur Fertilizer Policy for Lowland and Upland Rice Cropping Systems in Indonesia,* ACIAR Proceedings no. 29, 68–76. Canberra: Australian Centre for International Agricultural Research.

Amar, Boujemaa. 1987. "Ground Phosphate Rock as a Direct Application Fertilizer." In Indonesia, Departemen Pertanian, Pusat, Penelitian Tanah, *Prosiding Lokakarya Nasional Penggunaan Pupuk Fosfat, Cipanas, 29 June–2 July, 1987,* 463–488.

Barbier, Edward B. 1990. "The Farm-Level Economics of Soil Conservation: The Uplands of Java." *Land Economics* 66(2): 199–211.

Blair, Graeme, and Rod Lefloy, eds. 1990. *Sulfur Fertilizer Policy for Lowland and Upland Rice Cropping Systems in Indonesia.* ACIAR Proceedings no. 29. Canberra: Australian Centre for International Agricultural Research.

Booth, Anne. 1988. *Agricultural Development in Indonesia.* Sydney: Allen and Unwin.

Carson, Brian. 1989. "Soil Conservation Strategies for Upland Areas of Indonesia." Occasional paper no. 9, East-West Environment and Policy Institute, Honolulu.

Center for Soil and Agroclimate Research and International Soil Reference and Information Centre. 1994. "Soil Reference Profiles of Indonesia: Field and Analytical Data." Country Report 7. Draft. July.

Chin A Tam, S. M. 1993. *Bibliography of Soil Science in Indonesia, 1890–1963.* Haren, The Netherlands: DLO-Institute for Soil Fertility Research.

Diamond, Ray B., J. Sri Adiningsih, J. Prawirasumantri, and S. Partohardjono. 1988. "Response of Upland Crops to Water Soluble P and Phosphate Rock." *Prosiding Lokakarya Efisiensi Penggunaan Pupuk, Cipayung, 6–7 Agustus 1986* [Proceedings of the Conference on Efficiency in Fertilizer Application]. Bogor, Indonesia: Pusat Penelitian Tanah.

Diemont, W. H., A. C. Smiet, and Nurdin. 1991. "Re-Thinking Erosion on Java." *Netherlands Journal of Agricultural Science* 39: 213–24.

Donner, Wolf. 1987. *Land Use and Environment in Indonesia.* London: C. Hurst.

Geertz, Clifford. 1963. *Agricultural Involution: The Processes of Ecological Change in Indonesia.* Berkeley and Los Angeles: University of California Press.

Ismunadji, M. 1990. "N, P, K, and S Status of Food Crops in Islands Outside Java." In Graeme Blair and Rod Lefloy (eds.), *Sulfur Fertilizer Policy for Lowland and Upland Rice Cropping Systems in Indonesia.* Canberra: Australian Centre for International Agricultural Research. ACIAR Proceedings no. 29, pp. 83–86.

Lindert, Peter H. 1997. "A Half-Century of Soil Change in Indonesia." Working paper no. 90, Agricultural History Center, University of California–Davis, June.

Magrath, W. B., and P. L. Arens. 1987. "The Costs of Soil Erosion on Java: A Natural Resource Accounting Approach." Unpublished paper, World Resources Institute, Washington, D.C., November. Reissued as World Bank Environment Department working paper no. 18, 1989.

McCauley, David S. 1985. "Upland Cultivation Systems in Densely Populated Watersheds of the Humid Tropics. Opportunities and Constraints Relating to Soil Conservation: A Case Study from Java, Indonesia." Working paper, East-West Environment and Policy Institute, Honolulu.

Nibbering, Jan Willem. 1991. "Crisis and Resilience in Upland Land Use in Java." In Joan Hardjono, ed., *Indonesia: Resources, Ecology, and Environment,* 104–32. Oxford: Oxford University Press.

Repetto, Robert, William Magrath, Michael Wells, Christine Beer, and Fabrizio Rossini. 1989. *Wasting Assets: Natural Resources in the National Income Accounts*. Washington, D.C.: World Resources Institute.

Rosegrant, Mark W., and Faisal Kasryno. 1990. "The Impact of Fertilizer Subsidies and Rice Price Policy on Food Crop Production in Indonesia." In Graeme Blair and Rod Lefloy, eds., *Sulfur Fertilizer Policy for Lowland and Upland Rice Cropping Systems in Indonesia*, ACIAR Proceedings no. 29, 32–41. Canberra: Australian Centre for International Agricultural Research.

Santoso, Djoko, J. Sri Adiningsih, and Heryadi. 1990. "N, S, P and K Status of the Soils in the Islands outside Java." In Graeme Blair and Rod Lefloy, eds., *Sulfur Fertilizer Policy for Lowland and Upland Rice Cropping Systems in Indonesia*, ACIAR Proceedings no. 29, 77–82. Canberra: Australian Centre for International Agricultural Research.

Soemarwoto, Otto. 1991. "Human Ecology in Indonesia: The Search for Sustainability in Development." In Joan Hardjono, ed., *Indonesia: Resources, Ecology, and Environment*, 212–35. Oxford: Oxford University Press.

Strout, Alan M. 1983. "How Productive Are the Soils of Java?" *Bulletin of Indonesian Economic Studies* 19(1): 32–47.

Timmer, C. Peter. 1987. *The Corn Economy of Indonesia*. Ithaca: Cornell University Press.

van der Eng, Pierre. 1996. *Agricultural Growth in Indonesia since 1880: Productivity Change and the Impact of Government Policy*. Basingstoke, U.K.: Macmillan.

van Noordwijk, Meine, Carlos Cerri, Paul L. Woomer, Kusomo Nugroho, and Martial Bernoux. 1997. "Soil Carbon Dynamics in the Humid Tropical Forest Zone." *Geoderma* 79 (September): 187–225.

Index